The DevOps Career Handbook

The ultimate guide to pursuing a successful career in DevOps

John Knight

Nate Swenson

Packt>

BIRMINGHAM—MUMBAI

The DevOps Career Handbook

Copyright © 2022 Packt Publishing

Group Product Manager: Rahul Nair
Publishing Product Manager: Meeta Rajani
Senior Editor: Athikho Sapuni Rishana
Content Development Editor: Yasir Ali Khan
Technical Editor: Nithik Cheruvakodan
Copy Editor: Safis Editing
Project Coordinator: Shagun Saini
Proofreader: Safis Editing
Indexer: Sejal Dsilva
Production Designer: Jyoti Chauhan
Marketing Coordinator: Nimisha Dua

First published: May 2022

Production reference: 1110522

Published by Packt Publishing Ltd.
Livery Place
35 Livery Street
Birmingham
B3 2PB, UK.

ISBN 978-1-80323-094-8

www.packt.com

"In loving memory of Douglas Steven Swenson (1959–2022)

Without you and your unwavering support in my life, this book would have never been possible."

– Nate Swenson

Contributors

About the authors

John Knight is an engineering manager and former director with 17 years of DevOps experience, encompassing 8 different Fortune 500 companies. Lately, he focuses on digital transformations and enterprise migrations and wants to redefine what DevOps means for the next 10 years.

John is a lifelong learner, holding eight major cloud and DevOps certifications and two master's degrees, and he is currently working on a third, computer science, at Georgia Tech, where he hopes to focus on AI and ML and apply the intersection of the two fields to the next generation of AI/ML operations.

Outside of work, John spends time with his wife and four kids, enjoys streaming his favorite series, playing video games, and doing fun projects involving LEGO, robots, and Raspberry PIs.

Nate Swenson is a hands-on DevOps engineer with 12 years of experience within Fortune 100 companies in the insurance and finance sectors. Nate considers himself a DevOps generalist, although the majority of his work revolves around cloud infrastructure and **Continuous Integration (CI)** and **Continuous Delivery (CD)**.

Nate is the lead DevOps engineer for a team responsible for a self-hosted GitLab platform, which, in turn, is responsible for the source code management and CI/CD for several hundred applications. Nate specializes in cloud infrastructure, coaching, and CI/CD process improvements.

Outside of work, Nate can be found hanging out with his wife and two daughters, exploring the outdoors, or tinkering with his custom home automation setup when he is not working.

"I want to thank my wife for her unwavering support throughout the entire journey of writing this book."

About the reviewers

Daniele Fontani is the **Chief Technology Officer (CTO)** of Sintra Digital Business and has worked as a senior developer, team leader, and architect on a host of enterprise projects. He has a master's degree in robotic science and another master's degree in project management. His experience in technology extends to many technologies (Java, PHP, and .NET), platforms (SharePoint, Liferay, and Pimcore), and techniques (Agile, DevOps, and **Application Lifecyle Management (ALM)**. He is interested in agile techniques, project management, and product development. He implemented **Digital Experience Platforms (DXP)** for banks and the loan industry as a team leader and software architect. In the pharma industry, he has designed and developed retail portals for the training and social engagement of retailers.

> *"A special wish to all the readers of this book – may this book give you the tools to start a prosperous career."*

Satish Balakrishnan is an experienced cloud and DevOps architect with experience in companies such as Accenture, Hewlett Packard Enterprise, Singapore-MIT Alliance, and start-ups. He has an association with Microsoft as a Senior Cloud Solutions Architect for the APAC region. He has written numerous articles on the cloud and blockchain that have been published in the *CDOTrends, Techopedia, Blockchain Council*, and *IEEE* newsletters. Satish has also authored a book called *Terraforming the Cloud*.

Table of Contents

3

Specialized Skills for Advanced DevOps Practitioners

Section 2: The Application Process

4

Rebranding Yourself

5

Building Your Network

6

Mentorship

7
Working with Recruiters

Section 3: Interview Process

8
Preparing for Your Interview

9
Interviews Step by Step

Section 4: Tips, Tricks, and Interviews

10
DevOps Career: Tips and Tricks

11
Interviews with DevOps Practitioners

Index

Other Books You May Enjoy

Preface

Navigating any career is difficult; navigating one that is the culmination of many different careers can feel almost impossible. This book will aid you in navigating the field of DevOps and help prepare you for a career in it. The second half of the book focuses on techniques to use during each stage of the interview process to increase your odds of being a top candidate.

Who this book is for

This book is for anyone who wants to learn more about DevOps, pursue a career in DevOps, or advance their career in the field of DevOps.

What this book covers

Chapter 1, Career Paths, explores the history and culture associated with DevOps, followed by various career paths available in the field of DevOps.

Chapter 2, Essential Skills for a DevOps Practitioner, covers the skills required by all DevOps practitioners, regardless of level.

Chapter 3, Specialized Skills for Advanced DevOps Practitioners, covers the skills required for advanced careers within the field of DevOps.

Chapter 4, Rebranding Yourself, provides tips for updating your social presence as well as your résumé.

Chapter 5, Building Your Network, covers getting your skills noticed by the right people, which is key to landing a job, and offers tips and tricks on building your network to include the right people.

Chapter 6, Mentorship, focuses on the value of mentorship and how to connect with a mentor.

Chapter 7, Working with Recruiters, examines how most jobs are being filled by a combination of internal and external recruiters, giving you advice and tips on how to work with both.

Chapter 8, Preparing for Your Interview, provides tips on how to ensure you are prepared for your interview.

Chapter 9, Interviews Step by Step, walks you through what to expect at each stage of the process for both typical and non-typical interviews.

Chapter 10, DevOps Career: Tips and Tricks, provides a brain dump of the authors' 25 years of collective knowledge on things they have seen work and not work when interviewing.

Chapter 11, Interviews with DevOps Practitioners, revisits candid, open interviews with DevOps practitioners at various stages in their careers.

To get the most out of this book

This book assumes that you have a technical background. However, the only thing that is required to be successful in using this book is a desire to learn.

Basic software installed on your computer	Operating system requirements
Bash or Git Bash	Windows, macOS, or Linux
Nano or Vim	
Docker	

Download the color images

We also provide a PDF file that has color images of the screenshots and diagrams used in this book. You can download it here: `https://static.packt-cdn.com/downloads/9781803230948_ColorImages.pdf`.

Conventions used

There are a number of text conventions used throughout this book.

`Code in text`: Indicates code words in text, database table names, folder names, filenames, file extensions, pathnames, dummy URLs, user input, and Twitter handles. Here is an example: "The `-it` command runs the container with an interactive terminal."

Bold: Indicates a new term, an important word, or words that you see onscreen. For instance, words in menus or dialog boxes appear in **bold**. Here is an example: "Navigate back to GitLab to the project you just pushed, and click on **Settings | Pages** to view the URL where your site is published."

> **Tips or Important Notes**
> Appear like this.

Get in touch

Feedback from our readers is always welcome.

General feedback: If you have questions about any aspect of this book, email us at customercare@packtpub.com and mention the book title in the subject of your message.

Errata: Although we have taken every care to ensure the accuracy of our content, mistakes do happen. If you have found a mistake in this book, we would be grateful if you would report this to us. Please visit www.packtpub.com/support/errata and fill in the form.

Piracy: If you come across any illegal copies of our works in any form on the internet, we would be grateful if you would provide us with the location address or website name. Please contact us at copyright@packt.com with a link to the material.

If you are interested in becoming an author: If there is a topic that you have expertise in and you are interested in either writing or contributing to a book, please visit authors.packtpub.com.

Share Your Thoughts

Once you've read *The DevOps Career Handbook*, we'd love to hear your thoughts! Scan the QR code below to go straight to the Amazon review page for this book and share your feedback.

https://packt.link/r/1-803-23094-0

Your review is important to us and the tech community and will help us make sure we're delivering excellent quality content.

Section 1: A Career in DevOps

In this section, you will learn what it takes to be successful in DevOps, as well as various career paths and competencies required at various levels.

This section comprises the following chapters:

- *Chapter 1, Career Paths*
- *Chapter 2, Essential Skills for a DevOps Practitioner*
- *Chapter 3, Specialized Skills for Advanced DevOps Practitioners*

1
Career Paths

A DevOps career path is never linear, does not have a single point of entry, and can diverge at any moment. DevOps careers are rooted in Lean, Agile, and **Extreme Programming (XP)**, making it as much a culture fit between a candidate and employer as a technical fit. In this chapter, you will get a history lesson on DevOps that will aid in discussions with recruiters and hiring teams in the future. You will also be introduced to different skill profiles, which will help in determining the direction you take your career.

The following topics will be covered in this chapter:

- Why you should pursue a career in DevOps
- Overview of DevOps history
- DevOps culture
- DevOps career paths

Reading 200+ pages on a career you are not sure you want to pursue seems silly; time is a resource that is in short supply. In this section, we will cover why you should pursue a career in DevOps; specifically, why you should choose DevOps over other IT-related careers.

Earning potential

DevOps is constantly ranked as one of the highest-paying professions, with a median salary of $100,000. Entry-level DevOps engineers can expect to earn anywhere from $75,000 up to $145,000. As you progress in your career, you can expect to earn more. Look at the following graph:

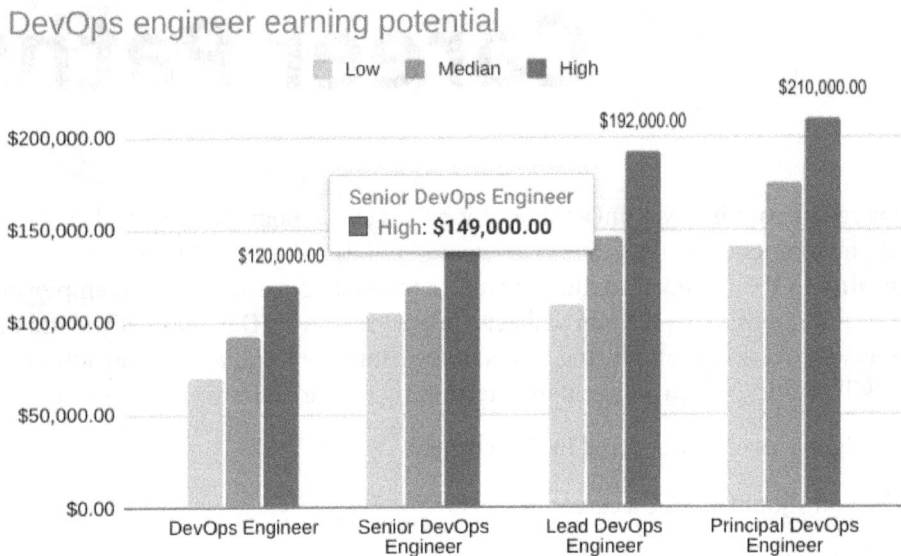

Figure 1.1 – DevOps salaries

Another reason you should consider a career in DevOps is you will never get bored as it is ever-evolving, which allows many opportunities to learn new skills.

Constant learning opportunities

Part of your job as a DevOps engineer is to stay up to date with the latest tools, technology, and trends that are occurring in the industry. DevOps engineers get paid to learn! It is one of my favorite things about my role as a DevOps engineer. As a DevOps engineer, you will ward off boredom while at the same time future-proofing your career.

Impact on the company

As a DevOps engineer, you will be delivering features used and felt by every part of the company. There is no other technical position where your efforts will have such a significant impact on the business.

Flexibility

As a DevOps engineer, you will have the flexibility to work where you want, when you want. Remote work has become increasingly accepted across the technology industry; DevOps teams were doing this before it was considered cool. Collaboration tools including Slack and Jira have made asynchronous work possible. What this means is you do not have to work the same hours as the rest of your team – at least not all the time.

> **So, Why Should You Pursue a Career in DevOps?**
>
> As a DevOps engineer, you will be highly compensated, constantly learn and apply cutting-edge technology to solve problems impacting the entire business, and have the flexibility to work where and when you want.

Overview of DevOps history

You are reading a book on DevOps, likely meaning you have a basic understanding of what DevOps is; if not, there's no need to worry – that will be covered as well. The history of DevOps is less known even within DevOps communities. First, we'll go back to understand key elements that came before DevOps that laid the groundwork and created the environment needed for DevOps to grow.

Lean manufacturing

Lean manufacturing is a production method aimed primarily at reducing cycle times within the production system as well as response times from suppliers and to customers.

The term **Lean** was coined in 1988 by *John Krafcik* and defined in 1996 by *James Womack* and *Daniel Jones*. Lean manufacturing is well-established as a set of best practices for manufacturing. Often branded as the *Toyota Manufacturing Method*, Lean manufacturing strives for process optimization across the manufacturing floor. *Continuous improvement* is the mantra for Lean manufacturing and practitioners continually evaluate ways to do the following:

- Keep inventory at a minimum.
- Minimize the queue of orders.
- Maximize efficiency in the manufacturing process.

Agile

In the early 2000s, traditional waterfall methods were evolving and being replaced by Agile, which required a large culture shift that focused on team empowerment. Agile is based around 4 core values and 12 principles. Some were adopted into DevOps as it evolved (`https://kissflow.com/project/agile/values-and-principles-of-agile-manifesto/`).

Extreme programming

XP aims to improve software quality and responsiveness to changing customer requirements. If you are thinking that sounds a lot like Agile, you wouldn't be wrong; it is a type of Agile software development. The biggest difference between XP and other Agile frameworks is the emphasis placed on the code and development (`https://en.wikipedia.org/wiki/Extreme_programming`).

The main contribution XP gave DevOps was **Continuous Integration (CI)**. CI was a term introduced in 2001 by *Grady Brooch* and was published as the Brooch method soon after.

DevOps

The exact inception of DevOps will forever be debated; it is widely accepted that between 2007 and 2008 is when the movement started. It was a perfect storm of events that allowed and triggered the DevOps movement. The dysfunction in the software industry, namely between IT operations and software development communities, was the spark that ignited the movement, but it was the pioneers of Agile, Lean, and XP who were responsible for the initial fuel of the DevOps movement.

In a world absent of DevOps, developers and IT operations belonged to different corporate hierarchies and **Key Performance Indicators** (**KPIs**) for IT operations and development were asynchronous and detrimental to the other. These conditions created teams siloed from one another, causing a breakdown in communication, and ultimately leading to failed deployments, missed deadlines, and angry customers.

In 2008 Andrew Shafer, a software engineer, tried to put together a meetup session entitled *Agile Infrastructure* at an Agile conference in Canada. Patrick Debois, an Agile practitioner, was the only one there. The two had a long conversation, which today is known as the spark that ignited a fire that became a movement known as **DevOps**. Andrew and Patrick formed a discussion group for other people to post their ideas for how to solve this divide between development and operations later that year. In 2009, the first DevOpsDays was held, in Belgium, which turned DevOps into a buzzword forever cemented in history. The DevOps movement continued with local meetups around the globe. Around 2010, open source software focused on DevOps began growing in popularity; Jenkins CI server software and Chef infrastructure provisioning software were a couple of pioneers.

> Pro Tip
>
> Understanding the history behind the job title you are applying for will make you seem more serious about the role and conversation much more natural. Dig deeper and read some books such as *The DevOps Handbook* and *The Phoenix Project*. They will only increase your chances of success further.

The following diagram gives a timeline of key dates in the history of DevOps:

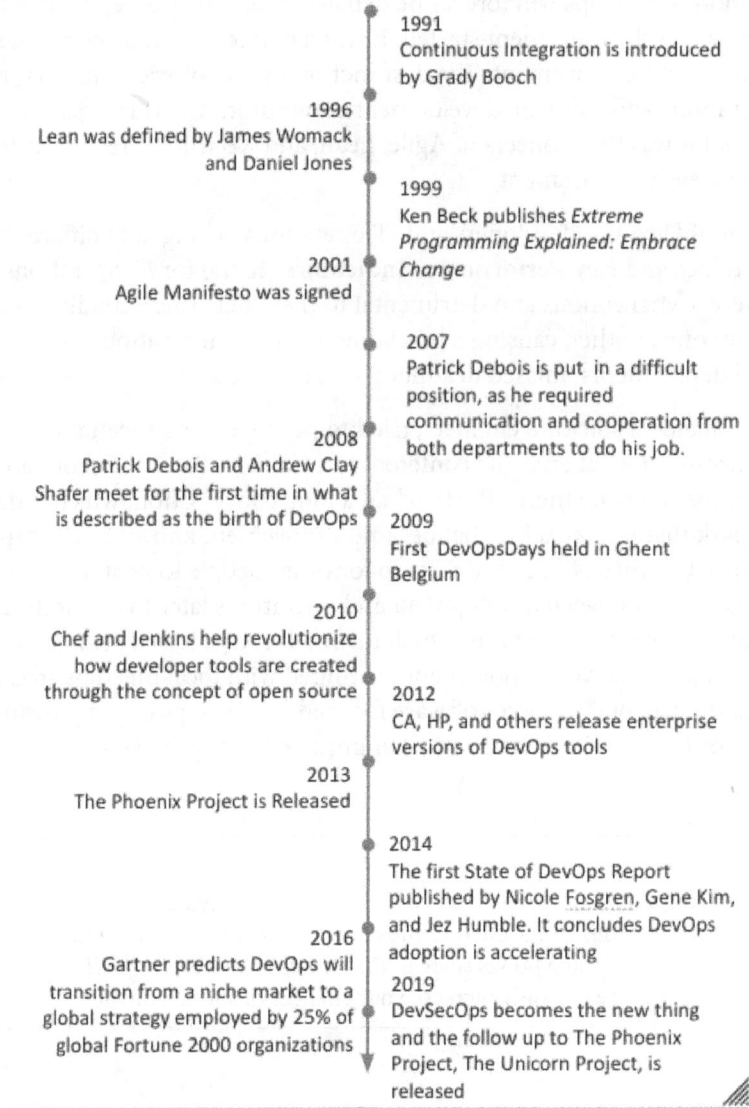

1991
Continuous Integration is introduced by Grady Booch

1996
Lean was defined by James Womack and Daniel Jones

1999
Ken Beck publishes *Extreme Programming Explained: Embrace Change*

2001
Agile Manifesto was signed

2007
Patrick Debois is put in a difficult position, as he required communication and cooperation from both departments to do his job.

2008
Patrick Debois and Andrew Clay Shafer meet for the first time in what is described as the birth of DevOps

2009
First DevOpsDays held in Ghent Belgium

2010
Chef and Jenkins help revolutionize how developer tools are created through the concept of open source

2012
CA, HP, and others release enterprise versions of DevOps tools

2013
The Phoenix Project is Released

2014
The first State of DevOps Report published by Nicole Fosgren, Gene Kim, and Jez Humble. It concludes DevOps adoption is accelerating

2016
Gartner predicts DevOps will transition from a niche market to a global strategy employed by 25% of global Fortune 2000 organizations

2019
DevSecOps becomes the new thing and the follow up to The Phoenix Project, The Unicorn Project, is released

Figure 1.2 – History of DevOps timeline

Now that we have learned about the history of DevOps, let's look at DevOps culture in the next section.

DevOps culture

DevOps is a set of practices that combines software development and IT operations. It aims to shorten the systems development life cycle and provide continuous delivery with high software quality. DevOps is complementary with Agile software development; several DevOps aspects came from the Agile methodology (`https://en.wikipedia.org/wiki/DevOps`). Diving deeper into that definition, we learn DevOps is a multi-faceted practice. DevOps has seven guiding principles that combine to form DevOps culture. DevOps culture aims to decrease cycle time, apply incremental changes, and create a more streamlined development process.

The following diagram gives a graphical representation of the seven principles of DevOps:

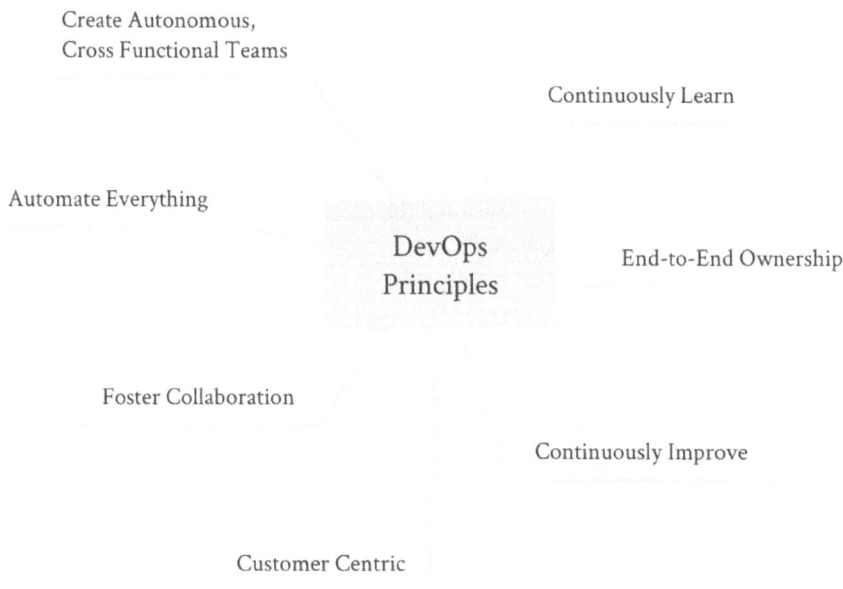

Create Autonomous,
Cross Functional Teams

Continuously Learn

Automate Everything

DevOps
Principles

End-to-End Ownership

Foster Collaboration

Continuously Improve

Customer Centric

Figure 1.3 – DevOps culture – principles

Now we will take a deeper dive into each of the seven principles of DevOps.

Customer-centric

Test often, get end user feedback frequently, and fail fast. The feedback loop between the customer and end users of products needs to be as short as possible. All actions taken by a team should be focused on the experience of the end user. This is also where the saying *shift-left* comes from, meaning *the sooner a feature is tested for bugs, the quicker it will be resolved*, and fewer downstream dependencies will be compromised.

A Tale of Two Start-Ups Bidding to Develop a Fitness Tracker for a Large Insurance Company

Need: A wearable device that will track users' fitness.

Acceptance Criteria: A wearable watch-like device that tracks various workouts.

Company X employs test-driven development and has weekly demos where they receive feedback. During a demo, they showcase a device and describe plans to track running and cycling. They learn that a big portion of their customers are swimmers. The team takes the feedback and shifts priority from running and cycling to swimming. The customer is impressed.

Company Y does not feel it is necessary to have demos as their past applications have done relatively well. The team focuses on the running and cycling workout tracking ability. During acceptance, they receive feedback that the watch must have the ability to track swimming. The development team is unable to meet the requirements in the given time frame. The customer is not impressed.

Outcome: Company X is awarded the contract and goes on to be a billion-dollar company. Company Y is not awarded the contract and receives poor press leading to another failed start-up.

Foster collaboration

The collaboration between the development team and IT operations teams is the most basic must for DevOps. Removal of silos ensures collaboration and alignment across entire organizations, ensuring a singular focus on the customer.

A collaborative culture is most effective when implemented using a top-down approach; executive sponsorship should be lined up ahead of any major culture shift. Another, much slower, approach is grassroots initiatives within an organization. A group of like-minded individuals with a platform to share on is all it takes to start a revolution. The trouble with the latter approach is overtime burnout can occur if you work tirelessly to make a change and see no results time and time again. Instead start with something you do have control over, such as your team.

Robert Weidner is a senior director at Optum and is one of only 26 Certified Enterprise Coaches in the world and is also my mentor and former manager. While working under Robert, our team was empowered to choose what micro team we worked in. We were also encouraged to hop over and help any other micro team who needed our support. When it came to stack ranking the team and fitting us to the bell curve for our bonus, we did our reviews of each team member during an offsite with the entire team present and in the hot seat while receiving feedback. It was frightening, but it worked because the team trusted each other.

End-to-end ownership

Feature teams ensure end-to-end responsibility by giving the team a vertical slice of a product, a feature. The feature, **Feature 1: Button X**, has two user stories: one for development and one for testing. The definition of done for the feature also requires the feature to be deployed successfully. This can be seen in *Figure 1.4*. The final piece to note is the ongoing support of *Button X* also remains with the team. Our company has started to call this **You Build It, You Own It (YBYO)**. The rationale behind this concept is that the team who built something is going to have the most knowledge about it when there is a production issue.

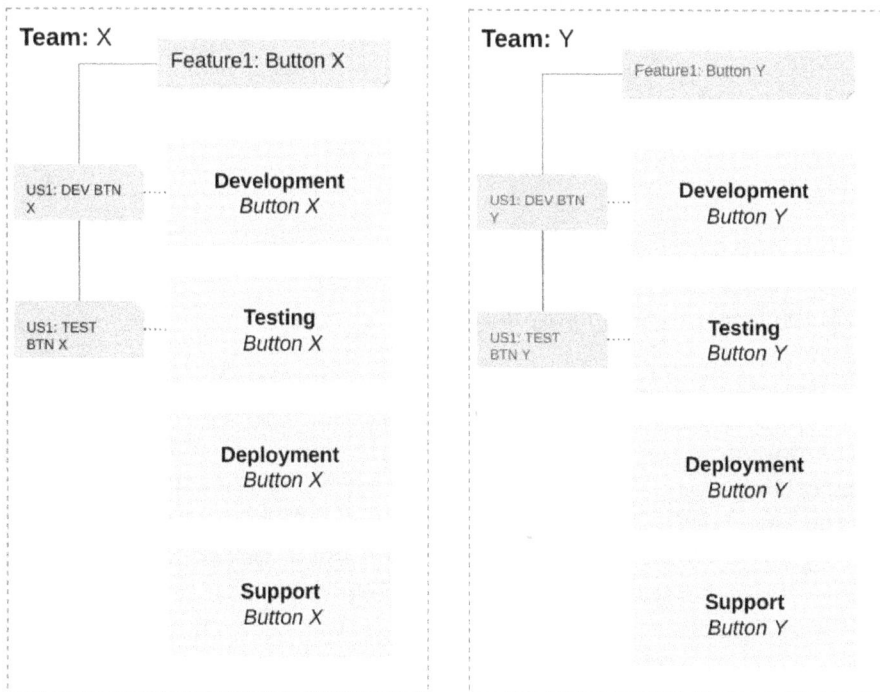

Figure 1.4 – Feature-centered team (E2E ownership)

In traditional development methods such as waterfall, teams are broken down and created at the activity level, also known as a horizontal slice of work. Ownership of a feature is split among various teams. In the following example, three teams must interact with the feature before it makes it to the end user, and another team is responsible for ongoing support. This is problematic; the operational support team is oftentimes not aware of the most recent changes the development team made, leading to extended downtimes and outages:

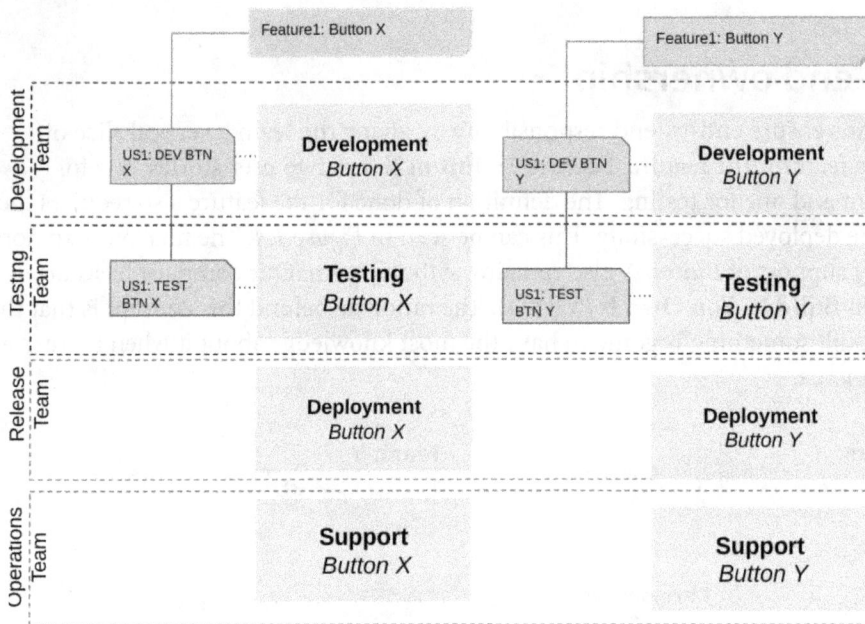

Figure 1.5 – Waterfall teams

Now, we'll talk about continuous improvement.

Continuous improvement

Continuous improvement was inherited from Lean. The entire team should be encouraged and, more importantly, empowered to make changes without fear of failure. Teams instead use failure as opportunities to improve on flawed processes. This is also known as **failing forward**. Failing forward allows for better control over risks as well as continuing to push the team forward. For your entertainment, the following is a script (`continous_improvement.sh`) to ensure your team is empowered to make improvements, continuously:

```
vi ci.sh                              ×    +                          _  □  ×

var COMPANY_EXISTS = true #Assuming your company exists

while[ $COMPANY_EXISTS]
  var empower_team=false
  do
    $empower_team=true
    make_a_change()
    if[$failure_occurs]
    then
      fix_issue()
        celebrate_lessons_learned()
    else
      celebrate_success()
done

make_a_change(){
  #do_something_amazing
}

fix_issue(){
  #do_something_to_fix_issue
}

celebrate_success(){
  echo "WOOHOO!!"
}
|
~
~
~
~
~
~
~
                                          28,0-1        Bot
```

Figure 1.6 – Continuous improvement shell script

The preceding script is a simple script that defines the flow of how the continuous improvement flow would operate if it were a shell script.

Automate everything

While doing research for this book, I noticed two common wordings: *automate everything* and *automate (almost) everything*. Further research revealed a common theme in types of processes that should not be automated, items with no payback, and items including a high degree of design or visual inspection, as seen in the following list (https://dzone.com/articles/what-to-automate-and-what-not-to-automate):

- Automation with no ROI
- Design
- Final QC of an application

Processes should have the least amount of manual intervention possible. The reason for this is simple: humans are error-prone while machines (computers) are excellent at executing high-volume repeatable tasks.

Next, we will cover continuous learning, a DevOps principle that is important for individuals looking to enter the field of DevOps, as well as those looking to stay relevant in the ever-changing field of technology.

Continuous learning

Technology is evolving at an astonishing rate; the most well-known example of this is Moore's law. Moore's law is the observation that the number of transistors in a dense **Integrated Circuit** (**IC**) doubles about every 2 years (`https://en.wikipedia.org/wiki/Moore%27s_law`). The number of transistors that fit into a microprocessor reached over 10 billion in 2017. It was under 10,000 in 1971(`https://ourworldindata.org/technological-progress`). Being a continuous learner is a personal attribute that will get you hired.

> **Pro Tip: You Must Be a Continuous Learner If You Wish to Succeed in DevOps**
>
> Creating a public project using a new technology is a great way to showcase this to potential hiring managers. Another way is to make sure to leave digital breadcrumbs of the most recent articles you have read, whether it be a post on LinkedIn or a tweet on Twitter.

An example that sticks out is an interview for a senior DevOps engineer role that was down to the final two candidates. Both candidates had tenure with the organization, exceeded the qualifications, interviewed well, and had advanced degrees. The candidate that received an offer displayed a hunger for knowledge throughout the interview process in subtle ways. The candidate chose to focus not on their degree but on a side project that had the purpose of teaching the candidate Golang. The theory of data science was being demonstrated with the application and it was cool. What stuck out, and continues to, was the candidate's desire to learn new things.

In summary, the combination of development and operation along with the seven DevOps principles, when applied together, form the DevOps culture. DevOps is a completely unique derivative of Lean, Agile, and XP aiming to shorten the feedback loop between development and the end user.

Take a look at the following visual depiction of DevOps culture broken down into the practices and principles:

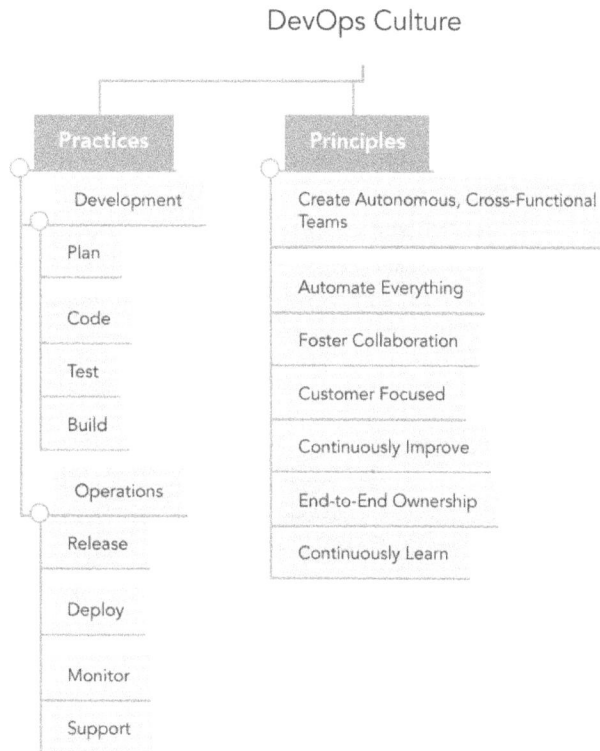

DevOps Culture

Practices	Principles
Development	Create Autonomous, Cross-Functional Teams
Plan	
	Automate Everything
Code	
	Foster Collaboration
Test	
	Customer Focused
Build	
	Continuously Improve
Operations	
	End-to-End Ownership
Release	
	Continuously Learn
Deploy	
Monitor	
Support	

Figure 1.7 – DevOps culture chart

In summary, DevOps culture contains seven guiding principles, as seen in *Figure 1.7*. In the next section, different career paths for a DevOps engineer will be discussed.

DevOps career paths

The field of DevOps is dense and is challenging to navigate, even for experienced practitioners. DevOps consists of eight core practices and follows seven basic principles. Unsurprisingly, there are numerous career paths in the field of DevOps. The generalist is the most common DevOps role.

A **DevOps generalist** is comparable to a Swiss Army knife, which is designed to handle as many tasks as possible. It can cut rope, open a can, cut wire, and if needed, the Swiss Army knife could fillet a fish. A DevOps generalist can create a deployment pipeline, write infrastructure as code scripts, and manage an **Elastic Kubernetes Service** (**EKS**) cluster in **Amazon Web Services** (**AWS**) if necessary.

A **DevOps specialist** is comparable to a fish fillet knife, which is singularly designed to fillet fish most effectively. The knife's profile, blade material, and ergonomics are finely tuned for a singular task, slicing fish. For example, a DevOps cloud specialist has spent their career focused on becoming an expert in cloud infrastructure, cloud architecture, and cloud security, and managing an EKS cluster in AWS is what they do in their sleep. It is likely that they would find a more cost-efficient way to do it than a non-cloud specialist.

A **DevOps specializing generalist** is comparable to an **Everyday Carry** (**EDC**) knife with a trailing point blade. This knife has ergonomics similar to a Swiss Army knife but a blade profile giving it the ability to fillet fish at a comparable level to a fillet knife. A DevOps specializing generalist who spent the past 10 years working in an AWS environment would be able to complete most DevOps tasks but would excel at those that involved AWS services.

Common skill profile shapes for a generalist, specialist, and specializing generalist can be seen in the following diagram:

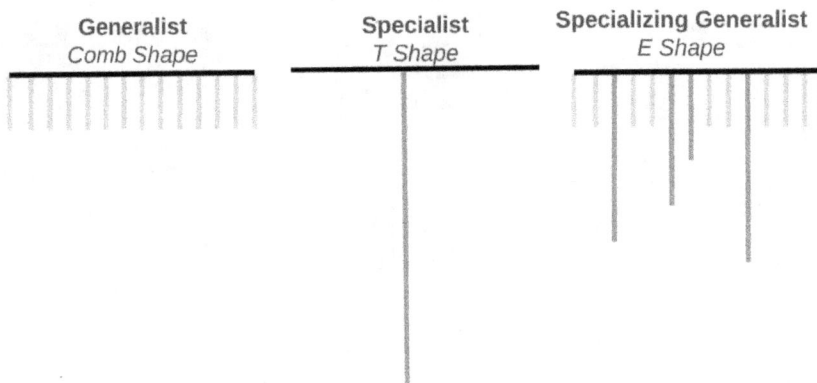

Figure 1.8 – Skill profiles

The profile of your skill set can be very useful when determining how to classify yourself. Start with a comb. Each prong (skill) has a similar length (depth). The comb shape is typical of a generalist. The second common profile is the T shape. A T has a single line (skill) that has a full length (depth). The T shape is typical of a specialist. An E-shaped profile, sometimes referred to as an unequal comb, has prongs (skills) of differing lengths (depths). Oftentimes, one or several skills have a definitively greater depth than the rest. The E shape is common for a specialized generalist.

A mentor told me the unequal comb (*E* shape) skill profile is the only true measure of an individual's skills. The comb shape is flawed because it assumes all skills have an equal depth, which is impossible. The *T* shape has a lack of detail; it shows a singular skill the individual is highly adept at but does not account for the other skills possessed by the individual.

> **Pro Tip**
>
> For this chapter, do not focus on what is required for the example skills listed as they will be covered in the next chapter. Instead, focus on how the skill profiles relate to the different types of DevOps engineers' specific requirements.

Common skill profiles fitted to the *E* profile can be seen in the following diagram:

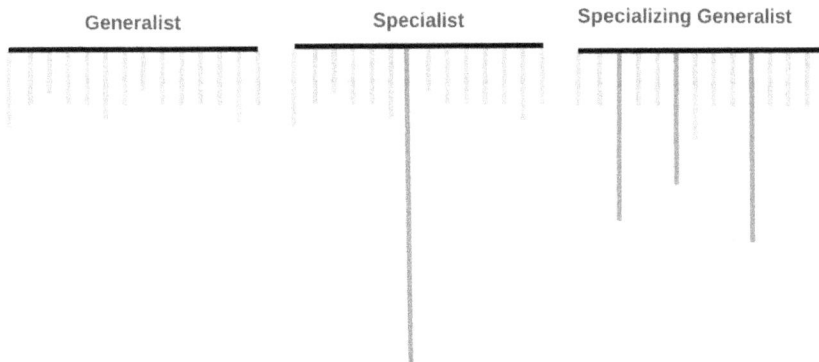

Figure 1.9 – Skill profiles (E profile fitted)

In the following sections, we will take a look at skill profiles for a DevOps generalist, specializing DevOps generalist, as well as several skill profiles for DevOps specialties. We will begin by looking at the skill profile for the DevOps generalist.

DevOps generalist

Google's Ben Fried stated, *Generalists, not specialists, will scale the web* (`https://devops.com/specialists-vs-generalists-enterprise-devops/`). This is a quote from back in 2011 but still holds true to an extent today. A generalist understands the entire **Software Development Life Cycle (SDLC)**. The generalist has broad knowledge across domains and skill areas but lacks a deep understanding of any domain. This is common in small companies, start-ups, or vertically integrated companies having only a handful of products and tools to support. There needs to be almost no handoff of work, leading to fewer places to drop the ball or forget something.

The following is a sample skill profile for a DevOps generalist. Take note of the relatively flat shape:

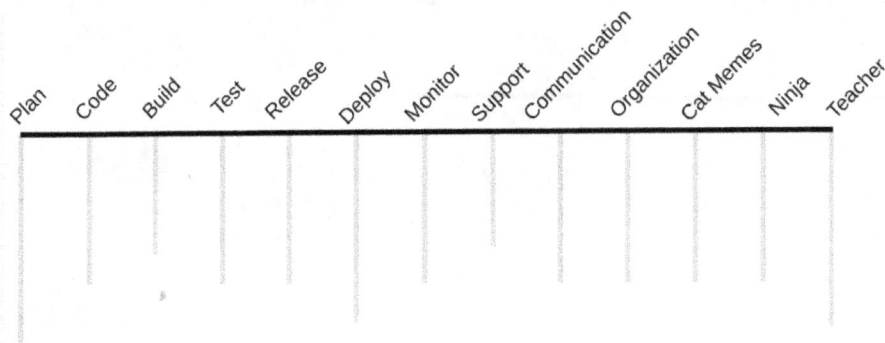

Figure 1.10 – DevOps generalist skill profile

The DevOps generalist is one of the first to feel the pains of growing or introducing new software, especially in smaller companies or organizations with a small DevOps department. When a new tool is introduced, the DevOps engineer needs to understand the tool before it is implemented. Next, an environment must be provisioned for the new tool, oftentimes unlike the environment the existing tool uses, or at least with quirks and subtle differences. Finally, the tool is implemented. At this point, both the new and existing tool need to be supported by the DevOps engineer. The comb shape that was used graphically to describe the generalist has an inherent flaw: it assumes a generalist does not have a deep understanding of any domain.

DevOps specializing generalist

If you stay working in the same industry or with the same type of projects long enough, you will eventually become a specializing generalist. The specialized generalist is also referred to as a master general. In the preceding example, the DevOps engineer has all the required skills but has a much deeper understanding and knowledge in the domain of programming and developing code. This is typical for software engineers who transition into a DevOps role. This could also be a skill profile you evolve over time if you enjoy certain skills, or if you just happen to always be assigned to tasks that require those skills. Regardless of how your skill profile evolved, knowing you have a deeper understanding of certain areas can be beneficial both when looking for a job as well as when it comes time to ask for a promotion.

In the following figure is the skill profile of a DevOps engineer who has knowledge in many domains with a deeper understanding of the **Build** and **Deploy** domains:

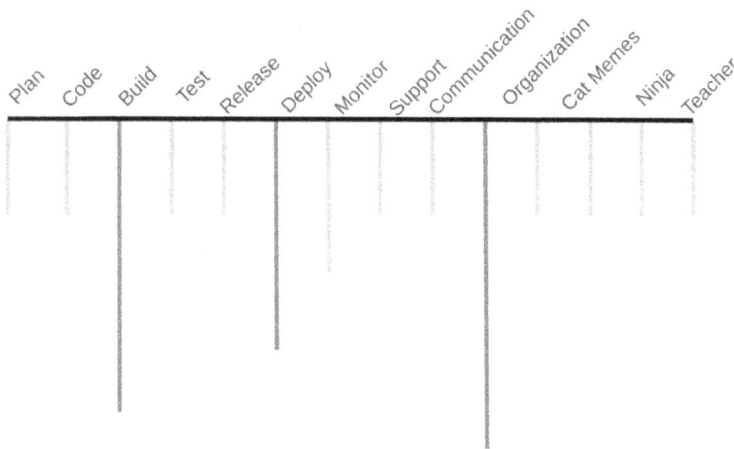

Figure 1.11 – DevOps specializing generalist skill profile

Understanding your own skill profile can be very helpful both for a manager of a team when planning capacity, as well as an engineer when choosing which roles to apply for, which is likely what you are interested in. In the next chapter, we will take a deeper dive into how you can create your own skills profile.

A specialist has a very deep understanding of a singular domain. This does not mean they are not capable of doing other things; however, it is usually not efficient to have a specialist work across domains. Specialists are more common in large organizations that define specialists by the tools they support. If we apply this to our previous example, DevOps engineering team *A* would specialize in tool *A*. When a new tool is introduced into the company, a new team would be formed and talent would be hired with correct skills or existing employees would be up-skilled to join the team. Specialists who possess deep knowledge in one of the DevOps domains are the ones that will be focused on in this book going forward.

DevOps security specialist

A DevOps engineer specializing in security is known to be in the niche field of DevSecOps. DevOps security specialists have a deep understanding of areas such as penetration testing, cloud security, chaos engineering, and continuous verification as seen in the following example skill profile:

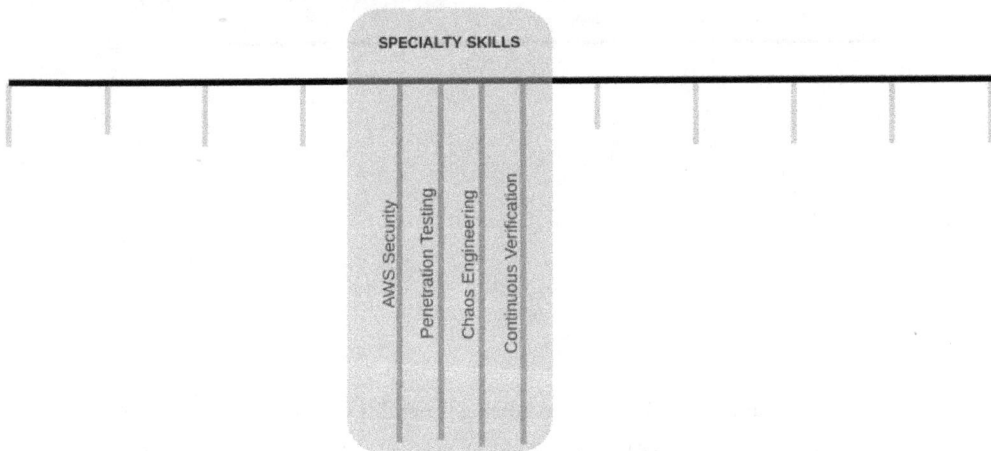

Figure 1.12 – DevOps security specialist skill profile

Now, we will talk about DevOps cloud specialists.

DevOps cloud specialists

One of the fastest-growing fields is cloud engineering and cloud engineers are some of the highest paid. What is the difference between a cloud specialist, a cloud engineer, and a DevOps cloud specialist? you might be asking, and the honest answer is nothing. Titles are meaningless and, oftentimes, they differ between companies and sometimes even between departments in larger organizations. A DevOps cloud specialist has traditional DevOps skills but has a very deep knowledge of the cloud tools, architectures, best practices, and management of entire cloud environments, sometimes multi-cloud environments.

The wide-scale adoption of the cloud has made a strong understanding of the cloud something even entry-level jobs oftentimes require.

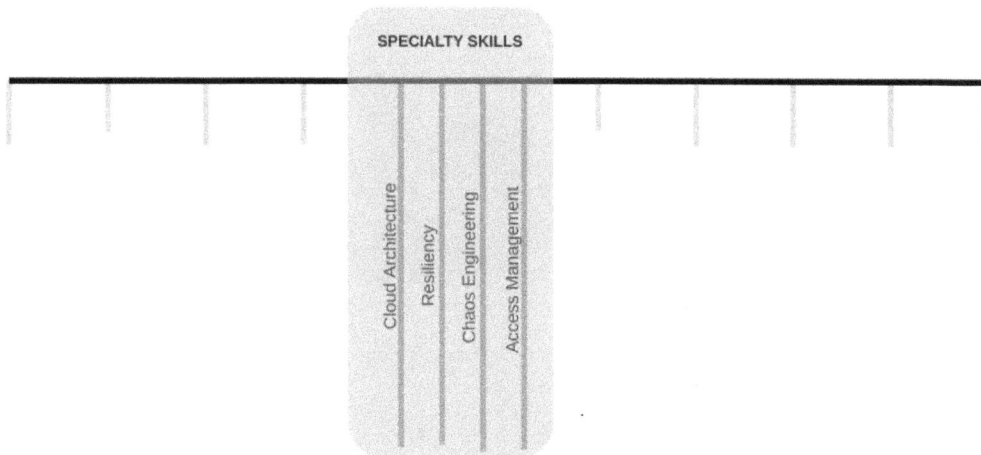

Figure 1.13 – DevOps cloud engineer specialist skill profile

In this section, skill profiles for DevOps generalists and DevOps specialists were covered.

Summary

In this chapter, you gained insight into the history and goals of DevOps. DevOps was founded around 2008 when a developer and Agilist met at a conference and decided there needed to be a better way of developing software. The goal of DevOps is to remove silos between development and IT operation teams as well as shortening the feedback loop between developers and customers. DevOps is a mix of Lean, Agile, and XP methodologies.

You also learned about the numerous career paths to choose from in the field of DevOps. Career paths in DevOps are defined by the depth of knowledge required. Three skill profiles were discussed: DevOps generalist, DevOps specializing generalist, and DevOps specialist. A DevOps specialist has a much greater depth of knowledge than a generalist.

In the next chapter, we will cover specific skills required for a DevOps generalist.

2
Essential Skills for a DevOps Practitioner

The most common question heard from individuals looking to land their first DevOps role is, *What are the important skills?* The list of important skills is rather lengthy for a DevOps generalist, the type of role most common for individuals looking for their first role in DevOps. This chapter is geared toward individuals looking for their first role in DevOps. Individuals who already have careers in DevOps can skip this chapter or use it as a refresher.

In this chapter, we will cover the following main topics:

- Scripting, coding, and programming
- Source code management
- Infrastructure management
- CI/CD concepts
- Soft skills
- Cloud-native frameworks
- Beginner DevOps certifications

Scripting, coding, and programming

There are DevOps engineers who are highly skilled at programming. However, you don't have to be a great programmer. To be a great DevOps engineer, debugging code, automation through scripting, and working in text-only terminals are also important skills. In this section, we will cover the following:

- Navigating the command line
- Scripting
- Modifying legacy code versus writing new code

Navigating the command line

Command-line navigation is possibly the most essential skill for anyone looking to get a job in the field of DevOps. Command line is a generic term that applies to a text-based interface. The most common command-line shell that comes with most Linux distributions is **Bash**. While having a basic knowledge of the usage of the command line is essential to land a job as a DevOps engineer, mastery of it can help you stand apart from other applicants. There is no way to master the terminal without making it a part of your regular daily routine. There are many well-written, resourceful blogs, articles, and cheat sheets that cover the command syntax. The focus of this section will be on techniques to quickly improve your comfort level when using the command line.

All navigation can be done through the command line and it is often much quicker. Even if you're not sure where you are, **path with directory (pwd)** has your back. pwd will output the current path you are at in the terminal. If you would like to see what files and folders are in the current directory, use the `ls` command. To navigate to a particular folder, use the `cd` command. If you want to find a particular string of text in a text file, you can use the `grep` command. `grep` is case sensitive by default, but like most commands, there are flags that can be applied to change its behavior; for example, `grep - i` makes the search insensitive to case. If there are multiple results, you can pipe your results with `sort` to sort the results alphabetically, or even in reverse order.

Figure 2.1 shows the Bash terminal window, where some basic commands have been executed and output can be seen:

```
 natejswenson@penguin:~/local_rep    ×    +                            _  □  ×

→  local_repo pwd
/home/natejswenson/local_repo
→  local_repo ls
backstage  shell_favorites
→  local_repo cd shell_favorites
→  shell_favorites git:(main) ✗ ls
help.sh  img  nav.sh  README.md  sites.csv  test.sh  todo.txt
→  shell_favorites git:(main) ✗ grep github sites.csv
 (github),folder
, (github),Profile,https://www.github.com
, (github),shell_bookmarks,https://github.com/natejswenson/shell_bookmarks
, (github),nateswenson.io,https://github.com/natejswenson/nateswenson.io
, (github),Profile,https://www.gitlab.com
→  shell_favorites git:(main) ✗ |
```

Figure 2.1 – Bash terminal with basic commands

As you can see in the preceding Bash terminal, there are some unique colors, the current directory is preceded by an arrow, and the branch that is checked out is shown for `git` folders. You can configure the appearance as well as setting aliases and custom functions within the `.bashrc` file.

> **Pro Tip: Play Around and Have Fun**
>
> Learning about the terminal should be fun, and we will expand on it in the next section. Till then, google commands, create your own, and be creative! The power of the terminal comes from its flexibility; the terminal can be customized to fit your needs and most of this customization is done through the `.bashrc` file.

The `.bashrc` file is the central area to set up aliases, functions, and customize the look and feel of your terminal. The `.bashrc` file is a shell script that loads when the terminal is loaded. If you are using Bash, you will have a `.bashrc` file, and a `.zshrc` file if you are using `zsh`. Within the `.bashrc` file, it is possible to do the following:

- Load modules:

```
module load <module>
```

- Modify an environment variable:

```
export PATH=$PATH:<path/to/dir>
```

- Activate a Python environment:

```
source <path/to/env>/bin/activate
```

- Set aliases:

Aliases are nicknames for commands, groups of commands, or scripts and can be added to the .bashrc file. Aliases are often created to make commonly used commands shorter. It is best practice to add aliases to a separate file called .bash_aliases, and then load .bash_aliases into .bashrc:

```
if [ -f ~/.bash_aliases ]; then
    . ~/.bash_aliases
fi
```

If your list of aliases is short, you can add them directly to the .bashrc file:

```
alias l="ls -l"
alias la="ls -la
```

Aliases can also be used to call functions:

```
alias dd=dockerdown()
dockerdown(){
  sudo docker rm -f $(sudo docker ps -a -q)
  sudo docker ps
  sudo docker rmi -f $(sudo docker images -q)
  sudo docker images
}
```

Another useful skill all DevOps engineers should have, and master, is the use of text editors within the terminal.

Text editors are command-line tools that allow you to directly edit files from the terminal window. Common flavors are vim, emacs, and nano. Most Linux distributions have vim installed by default. In the following example, we will show you how to edit your .bashrc file. To open your file in vi, type vi </path/to/file>. In the following example, the command was sudo vi .zshrc, which opens the .zshrc file in vim with sudo privileges:

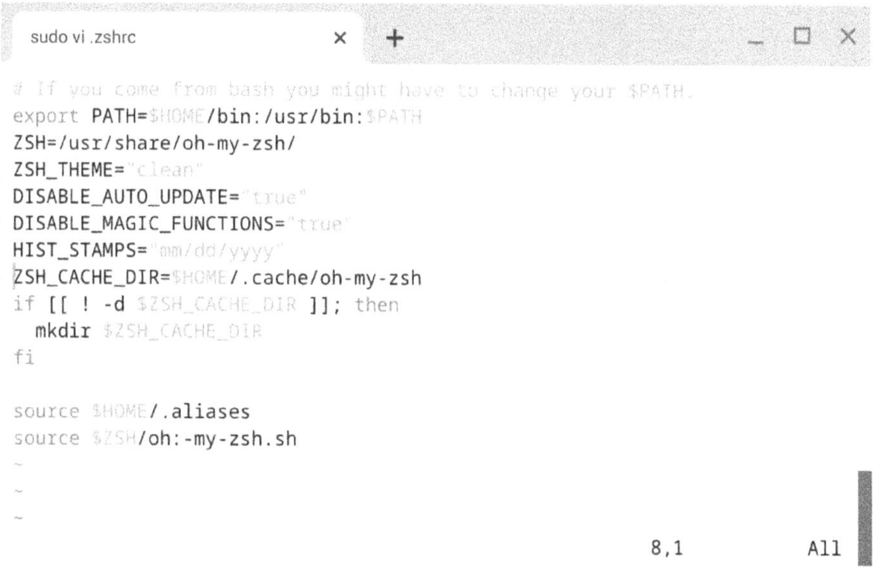

```
# If you come from bash you might have to change your $PATH.
export PATH=$HOME/bin:/usr/bin:$PATH
ZSH=/usr/share/oh-my-zsh/
ZSH_THEME="clean"
DISABLE_AUTO_UPDATE="true"
DISABLE_MAGIC_FUNCTIONS="true"
HIST_STAMPS="mm/dd/yyyy"
ZSH_CACHE_DIR=$HOME/.cache/oh-my-zsh
if [[ ! -d $ZSH_CACHE_DIR ]]; then
  mkdir $ZSH_CACHE_DIR
fi

source $HOME/.aliases
source $ZSH/oh:-my-zsh.sh
~
~
~
~
                                          8,1              All
```

Figure 2.2 – Bash terminal with the vi editor open

At this point, the file is opened in read-only mode. To enter edit mode, type *I*. To make your change, enter *esc* followed by *w* to save the file and *q* to close the editor.

Text editors are powerful but require some time to be mastered. If you need information on `vi` commands, you can find great resources on various online forums. A particular favorite of mine is `https://ryanstutorials.net/linuxtutorial/cheatsheetvi.php`.

Scripting

Scripts are something DevOps engineers must be capable of creating and maintaining. The secret to getting a DevOps job is being able to solve scripting problems, which means practice. There are several scripting languages currently used by DevOps engineers. No one of them is better than another; instead, each of them is best suited for distinct types of jobs.

> **Pro Tip: Google Is Your Best Friend when Learning a New Language**
>
> If you are struggling with something, chances are someone else has already struggled with it, solved it, and written about it. Don't work harder; instead, work smarter and more efficiently.

Python

Python is heavily used in infrastructure automation and provisioning and has become an all-purpose scripting language in DevOps. It is favored by many because it is easy to get started with. However, it gets exponentially more difficult as your proficiency progresses. The following is the most basic Python script (`helloworld.py`):

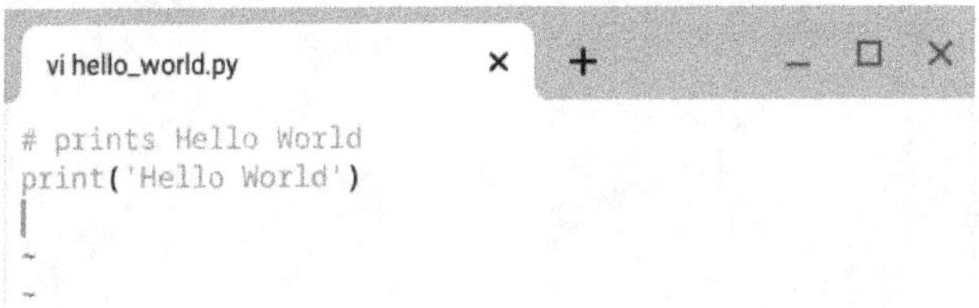

```
vi hello_world.py                      ×    +        _  □  ×

# prints Hello World
print('Hello World')
|
~
~
```

Figure 2.3 – hello_world.py

Bash

Bash is the most used scripting language in the Unix/Linux environment and has a strong community that provides support. It is used to automate Linux servers around the world. The following is the most basic shell script (`helloworld.sh`):

```
vi hello_world.sh                      ×    +        _  □  ×

#!/bin/bash

# print hello world

echo "Hello World" |
~
~
```

Figure 2.4 – hello_world.sh

JavaScript

JavaScript is used as DevOps scripting to create network-centric applications. It is a lightweight DevOps scripting Language. JavaScript offers numerous advantages, including less server interaction, increased interactivity, immediate feedback to visitors, and richer interfaces. The following is the most basic JavaScript script (`helloworld.js`):

```
vi hello_world.js                      ×    +        _  □  ×

// the hello world program

console.log('Hello World'); |
~
~
```

Figure 2.5 – hello_world.sh

Go

Go was introduced in 2009 and has drastically changed the DevOps landscape since its inception. Built on C, Go was created to be readable by humans and scalable. The following is the most basic Go script:

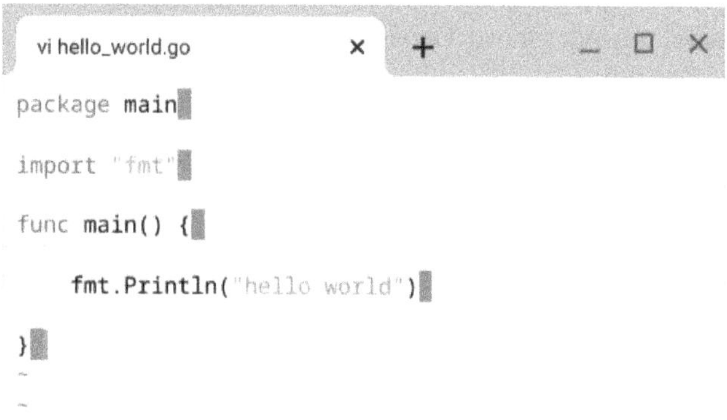

```
vi hello_world.go                    ×    +              —  □  ×

package main

import "fmt"

func main() {

    fmt.Println("hello world")

}
~
~
```

Figure 2.6 – hello_world.go

Pro Tip: Focus on Learning One Language at a Time

You set yourself up for disappointment if you try to learn multiple coding languages simultaneously. Unless circumstances require you to learn a new language, get proficiency in one language before moving on to the next.

There are tons of books and online resources you can choose from when learning new languages. A great way to practice is forking a project from GitHub and making changes to it. This is some of the most useful experience you can give yourself. If you would like to challenge yourself, you can try online sites designed specifically to prepare users for technical interviews, also offering a great way to upskill your coding game. Here are a couple of favorites:

LeetCode: `https://leetcode.com/`

AlgoExpert: `https://www.algoexpert.io/`

Other sites built around the type of problems you may see in an interview exist. Coding challenge sites can also help tremendously to increase your chances of success.

Now that we have covered various scripting languages, we need to cover when to modify existing code and when to write new code.

In this section, you learned about navigating text-based shells such as Bash, as well as how to modify existing files and create files using text editors within a shell. We also covered various scripting languages used by DevOps engineers and the best use cases for each.

In the next section, we will cover version control and source code management.

Source code management

Source Code Management (**SCM**) is the tool used to manage your code. Before we discuss SCM, it is crucial to understand **Version Control** (**VC**), which is the process used to manage your code. For this book, we will assume VC is synonymous with `git`.

Git

An astounding 87% of developers use `git` as their version control. Git is a distributed version control software initially designed by Linus Torvalds to manage the Linux kernel. The difference between `git` and `svn` is that the complete code history is stored on each individual node when using `git` versus a single source server when using `svn`. There are several things to consider when learning `git`:

- **First**: Git is available on almost every operating system – macOS, Windows, and Linux all have `git` versions available.

> **Getting Started with Git**
>
> Windows: `https://gitforwindows.org/`
>
> macOS: `https://git-scm.com/download/mac`
>
> Linux: `https://git-scm.com/download/linux`

- **Second**: There are multiple branching strategies. It is recommended that you spend time learning and practice managing your own projects using various strategies. Some common strategies are defined as follows:

Figure 2.7 – Basic git workflow

The basic `git` workflow has one branch, the main or *Master* branch. Developers commit directly to this branch and all deployments, regardless of the environment, are made from this branch. This is a workflow that is not recommended unless you need to get set up quickly or are working on a private side project.

The following diagram is a graphical representation of the git feature branch workflow:

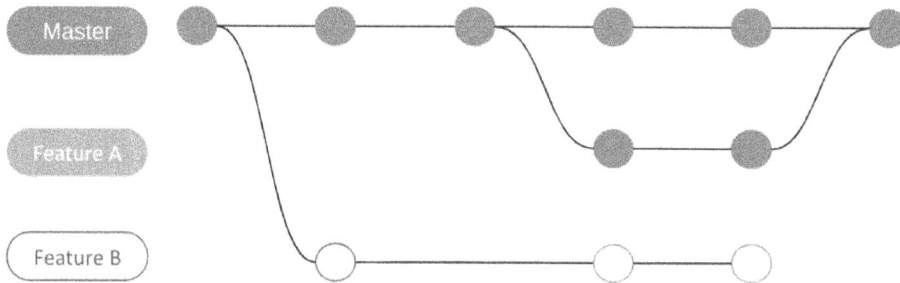

Figure 2.8 – Git feature branch workflow

The git feature branch workflow becomes necessary whenever there is more than one person working from the same code base. Both feature *A* and feature *B* can be created without the worry of affecting the other's ability to merge back to the *Master* branch.

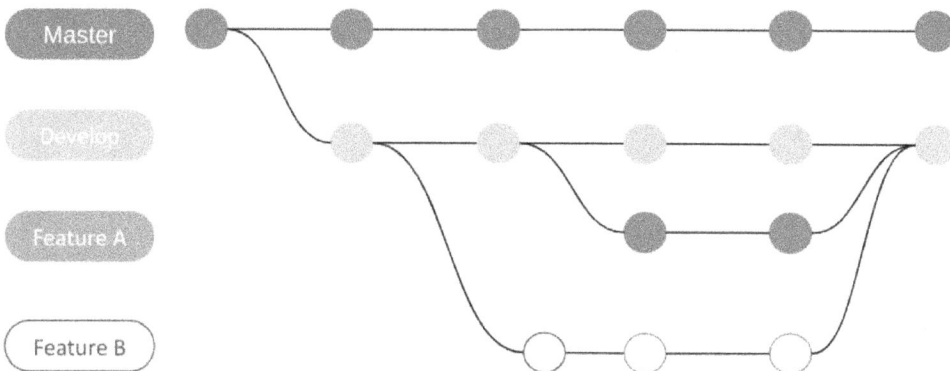

Figure 2.9 – Git feature branch workflow with a Develop branch

The git feature workflow with the *Develop* branch is one of the most popular branching strategies. The *Master* branch is always in a state that is ready to be deployed to production, and developers can work on their own features without worrying about merge conflicts from other developers.

- **Third**: There are a lot of git commands and there is no way to memorize them all. We have included two approaches to help you feel more confident when starting out.

Add commonly used `git` commands as aliases to your `.bash_profile` file. The following is a snippet of code that can be added to your `.bashrc` file that combines three commands associated with `git` into a single gp alias:

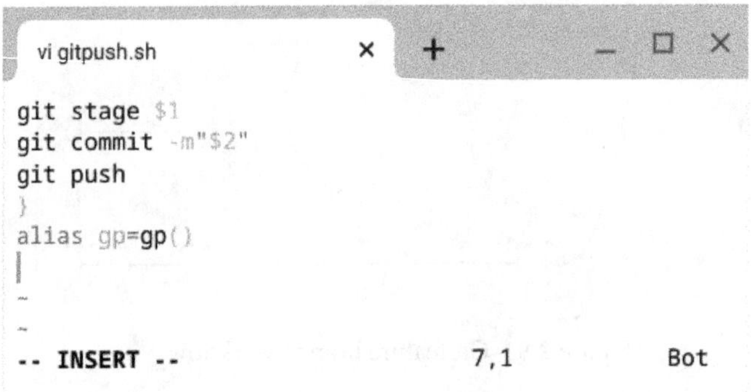

```
vi gitpush.sh                          ×   +              —   □   ×

git stage $1
git commit -m"$2"
git push
}
alias gp=gp()

~

~

-- INSERT --                            7,1              Bot
```

Figure 2.10 – Git function example for .bashrc

The preceding gp alias takes two parameters: `$1=file,` or the path of files to the stage, and `$2=commit message`. The following output shows what is seen when you execute the gp alias. Let's break this down:

- `shell_favorites` is a local working directory tracked by `git`.

- The `git stage` command moves `README.md` into the local staging area.

- The `git commit -m` command commits the `README.md` file to the local repository with a commit message, which is `test100721` in our example.

- The `git push` command pushes the changes that are in your local repository to the remote repository that `shell_favorites` is tracked to.

In the following figure, you can see the output the gp alias would result in:

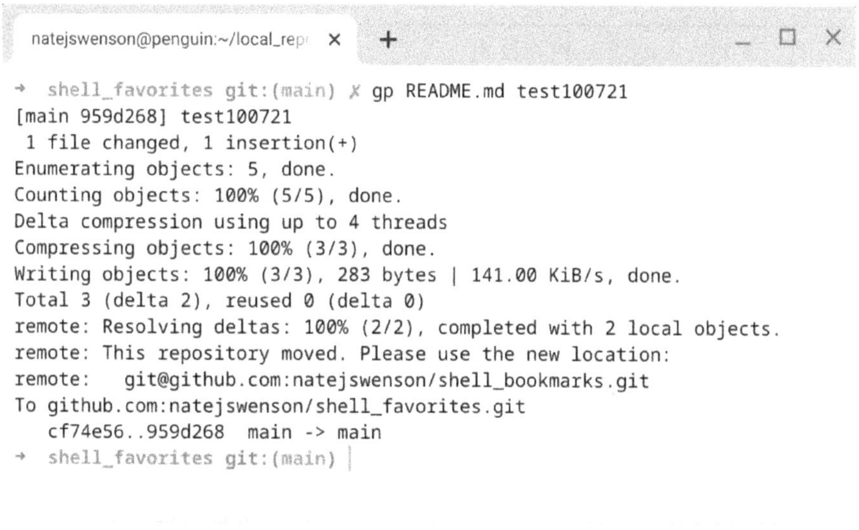

```
natejswenson@penguin:~/local_rep   ✕   +                              _   ☐   ✕

→  shell_favorites git:(main) ✗ gp README.md test100721
[main 959d268] test100721
 1 file changed, 1 insertion(+)
Enumerating objects: 5, done.
Counting objects: 100% (5/5), done.
Delta compression using up to 4 threads
Compressing objects: 100% (3/3), done.
Writing objects: 100% (3/3), 283 bytes | 141.00 KiB/s, done.
Total 3 (delta 2), reused 0 (delta 0)
remote: Resolving deltas: 100% (2/2), completed with 2 local objects.
remote: This repository moved. Please use the new location:
remote:   git@github.com:natejswenson/shell_bookmarks.git
To github.com:natejswenson/shell_favorites.git
   cf74e56..959d268  main -> main
→  shell_favorites git:(main)
```

Figure 2.11 – git push terminal output

Pretty sweet, no? Adding git aliases won't make you a better developer but it can simplify your life.

My next secret to success with git for beginners is having a list of commonly used git commands close at hand at all times – a git cheat sheet. A favorite of mine is the one by the education group of GitHub: https://education.github.com/git-cheat-sheet-education.pdf.

SCM

Popular SCM tools include GitHub, GitLab, and Bitbucket.

GitHub: https://www.github.com

GitLab: https://about.gitlab.com/

Bitbucket: https://bitbucket.org/

Each of these SCM tools has unique features to help improve developers' experiences and are designed to be user-friendly and easy to use. The reason these solutions tend to be easy to use is the rich UI each has developed for the user. You can use any of these tools for free by signing up on their websites! Regardless of the SCM tool you choose, the version control is still git.

In the following table, we will compare the three most popular SCM tools available as of 2021:

	GitHub	GitLab	Bitbucket
Free Public Repositories	x	x	x
Free Private Repositories	x	x	x
Merge and Issue Templates	x	x	-
Integrated CI	-	x	x
Open Source	-	x	-
Integrations	-	-	-

Table 2.1 – SCM comparison

At the end of the day, there really is no bad choice as long as you are learning.

In this section, you learned about `git`, common `git` patterns, and common `git` commands. We also discussed options available for source code management software.

In the next section, you will learn about infrastructure tools and techniques needed to be successful as a DevOps engineer.

Infrastructure management

Gartner defines IT infrastructure this way:

> *IT infrastructure is the system of hardware, software, facilities, and service components that support the delivery of business systems and IT-enabled processes.*

Infrastructure management can be broken down into three key stages: *capacity planning*, *infrastructure provisioning*, and *deployment*, as can be seen in the following diagram:

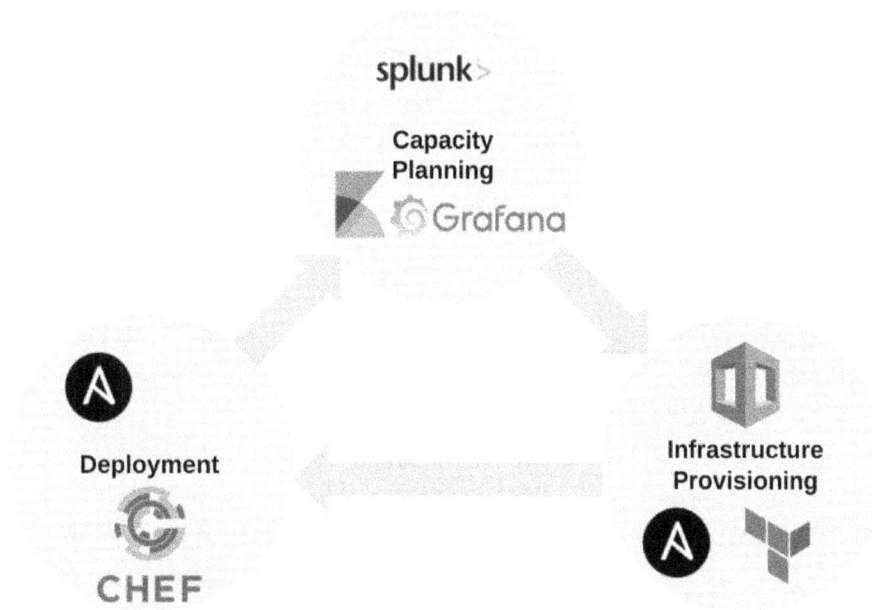

Figure 2.12 – Infrastructure management stages

The first stage in infrastructure management is capacity planning, which will be covered next.

Capacity planning

Capacity planning is the first step in infrastructure management and is followed by provisioning and deployment. Several tools can help with collecting the data needed to accurately plan resources as collecting accurate data is crucial to do so. Tools used during capacity planning are Splunk, the **ELK (Elasticsearch, Logstash, Kibana)** Stack, and New Relic. Continuous capacity planning is required to scale up and scale down resources based on demand in production.

Auto-scaling resources have two benefits, cost savings and better performance. When a resource is scaled down, it is removed from usage, which means it no longer incurs charges. When resources are scaled up, additional resources are added before performance degradation occurs.

Infrastructure provisioning

After capacity numbers are successfully collected and analyzed, we can move on to infrastructure provisioning. Provisioning involves the creation, allocation, and deletion of infrastructure resources based on the derived capacity numbers from capacity planning. Infrastructure resources include servers, containers, storage, networks, IPs, and load balancers that can be created and managed on cloud service providers such as AWS, Azure, GCP, or on-premises.

A DevOps engineer needs to know how to manage infrastructure resources in the cloud and on-premises environments depending on the company's architecture. In the following examples, you will be shown boilerplate code for creating AWS EC2 instances using CloudFormation, Terraform, and Ansible.

If the organization is managing resources on AWS, then AWS CloudFormation may be used to automate the creation/allocation/deletion of infrastructure resources. The following is a boiler template for CloudFormation used to provision an EC2 instance:

```yaml
vi ec2_cloudformation.yaml

Type: AWS::EC2::Instance
Properties:
  AvailabilityZone: String
  DisableApiTermination: Boolean
  HostId: String
  HostResourceGroupArn: String
  IamInstanceProfile: String
  ImageId: String
  KeyName: String
  LaunchTemplate:
    LaunchTemplateSpecification
  Monitoring: Boolean
  SecurityGroups:
    - String
  SubnetId: String
  Tags:
    - Tag
```

Figure 2.13 – CloudFormation example

To learn more about CloudFormation, visit `https://docs.aws.amazon.com/`
`AWSCloudFormation/latest/UserGuide/Welcome.html`. If your organization
is managing resources across multiple cloud services, providers such as AWS, Azure, GCP,
and Terraform can be used to automate the creation/allocation/deletion of infrastructure
resources. The following is a Terraform file that could be used to provide an EC2 instance:

```
vi ec2_terraform.tf                    ×    +                              _  □  ×

terraform {
  required_providers {
    aws = {
      source  = "hashicorp/aws"
      version = "~> 3.27"
    }
  }

  required_version = ">= 0.14.9"
}

provider "aws" {
  profile = var.profile
  region  = var.region
}

resource "aws_instance" "app_server" {
  ami           = var.ami
  instance_type = var.ec2_type
}

~
~
~
~
~
~
~
~
~
~
-- INSERT --                                      19,31          All
```

Figure 2.14 – Terraform example

Ansible can also be used to provide resources spread across various environments both on-premises and in the cloud. The following example will create an EC2 instance with the following variables passed in: MY_KEY, EC2_TYPE, IMAGE, GROUP, COUNT, and VPC_SUBNET:

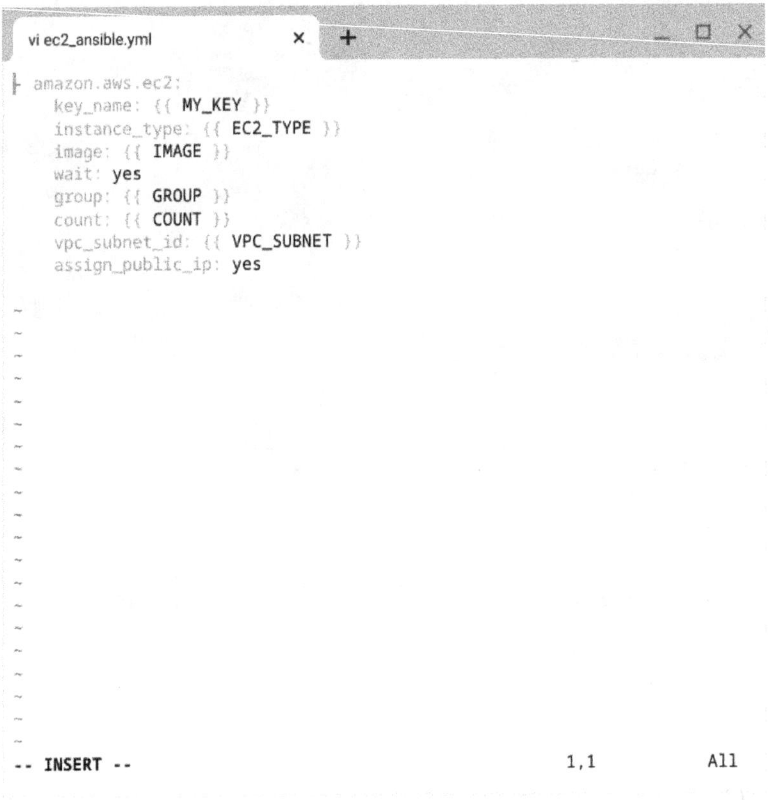

```
vi ec2_ansible.yml                        ×    +                                    _  □  ×

⊢ amazon.aws.ec2:
    key_name: {{ MY_KEY }}
    instance_type: {{ EC2_TYPE }}
    image: {{ IMAGE }}
    wait: yes
    group: {{ GROUP }}
    count: {{ COUNT }}
    vpc_subnet_id: {{ VPC_SUBNET }}
    assign_public_ip: yes

~
~
~
~
~
~
~
~
~
~
~
~
~
~
~
~
~
-- INSERT --                                              1,1          All
```

Figure 2.15 – Ansible example

Now, we will talk about deployment.

Deployment

After infrastructure has been provisioned, proceed to the deployment stage. Deployment involves installing, configuring, releasing, and managing software services on the servers or containers that serve the production workload. Deployment is a process that occurs within the servers or containers that are created or allocated during the automated provisioning of infrastructure resources. A DevOps engineer can use automation tools such as Chef, Ansible, and Salt to automate the deployment of software services.

CI/CD concepts

Continuous Integration (CI) and **Continuous Delivery (CD)** are synonymous with DevOps. This is because every practice discussed in *Chapter 1*, *Career Paths* – plan, code, build, test, release, deploy, and operate – is included in the infinite CI/CD loop, as shown in *Figure 2.16*:

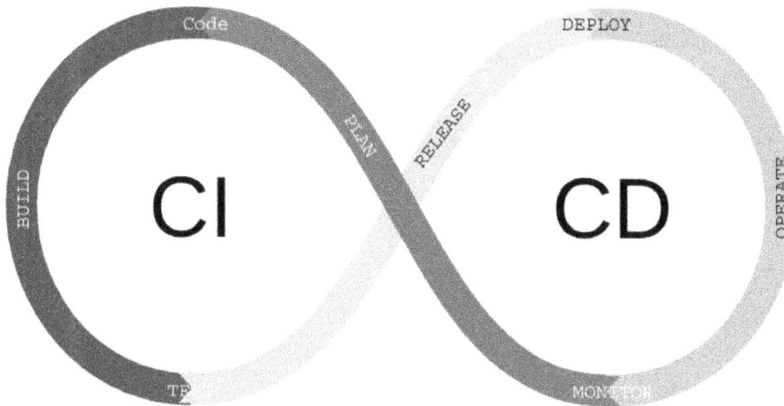

Figure 2.16 – Infinite CI/CD loop

Let's first examine CI and the related practices and tools associated with it.

Continuous integration

Continuous integration is the process of merging code changes from multiple developers into a single branch on a regular and frequent basis. To do this effectively, you need some form of automation that builds your code and executes a battery of tests against it. CI servers help to effectively integrate your code using CI pipelines.

After a developer makes a change, code changes are committed to a source code management system by using `git`. The CI server has a built-in listener (hook) to trigger a build whenever code is committed. The pipeline creates a new build and runs a battery of tests against the build. The tests include static code analysis, dynamic code analysis, secret detection, and vulnerability scans, as well as functional and integration tests. The following figure shows how a CI server interacts with several aspects of the development life cycle:

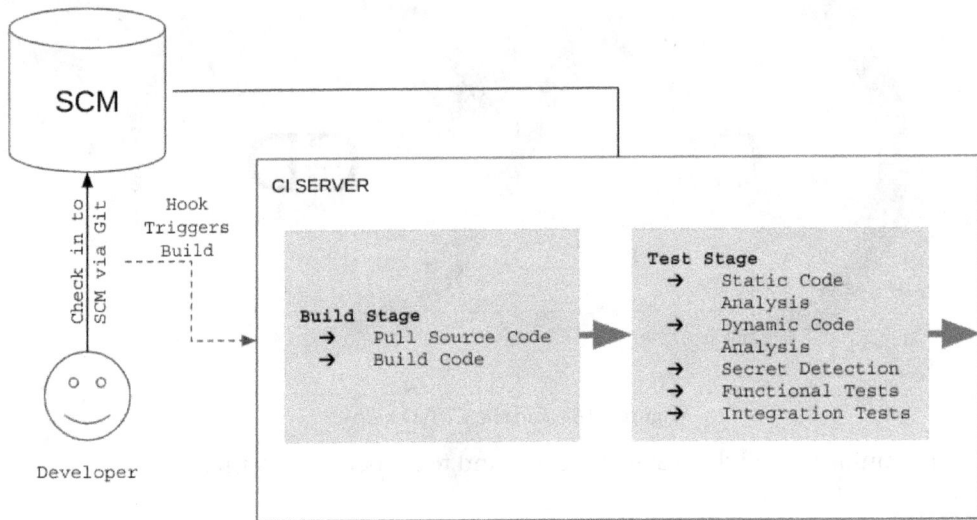

Figure 2.17 – CI pipeline

Continuous integration servers include Jenkins, **Travis CI**, **CircleCI**, and **GitLab**.

Each of these offers similar functionality with slight differences in the user interface and the language required to write the pipeline.

Jenkins

The first CI software we will discuss is the most widely used, Jenkins. Jenkins has a large community and many features due to it being open source, which is why it is also free to use. Some drawbacks to Jenkins are the overhead to maintain it as well as a complex pipeline design. Jenkins pipelines use **Groovy**, which is an offshoot of Java. The following is the controller-agent architecture used with Jenkins CI:

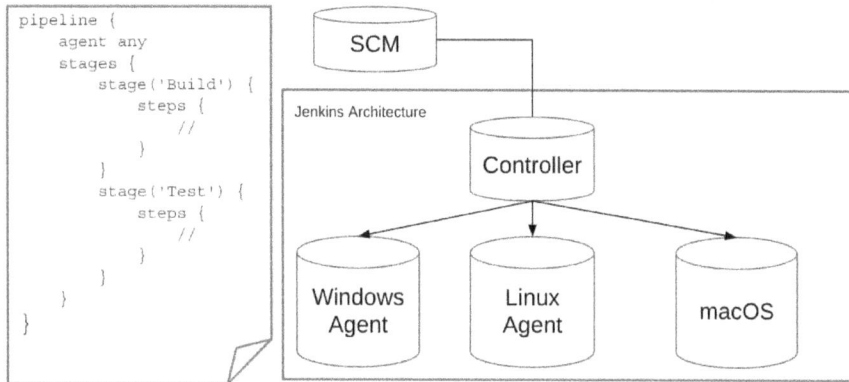

```
pipeline {
    agent any
    stages {
        stage('Build') {
            steps {
                //
            }
        }
        stage('Test') {
            steps {
                //
            }
        }
    }
}
```

Figure 2.18 – Jenkins architecture

GitLab

GitLab is both an SCM and CI tool. GitLab CI is new as of 2014, but its user base has grown exponentially since its release. GitLab CI uses the runner concept, which means each job runs in its own container-based executor. It offers a wide range of security tools. It can be difficult to manage if you run it on-premises. You can see in the following diagram that there are many servers that need to be managed and configured. However, GitLab also offers a **Software as a Service (SaaS)** option for smaller companies looking to get started more quickly. GitLab is YML-based, making it quite easy to write and understand pipelines. The following architecture diagram is one viable option to implement GitLab within an organization:

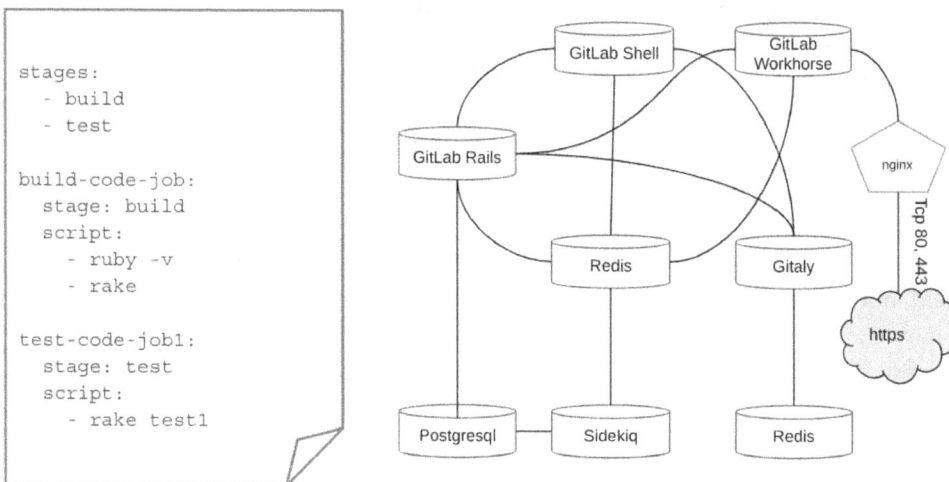

```
stages:
    - build
    - test

build-code-job:
    stage: build
    script:
        - ruby -v
        - rake

test-code-job1:
    stage: test
    script:
        - rake test1
```

Figure 2.19 – GitLab architecture

Continuous delivery

Continuous delivery is an extension of continuous integration. After the build stage, code changes are delivered to higher environments such as stage, test, preprod, and prod. With continuous delivery, an automated release process must be in place as well. The following diagram shows the CI server stages identified as being a CI or CD task:

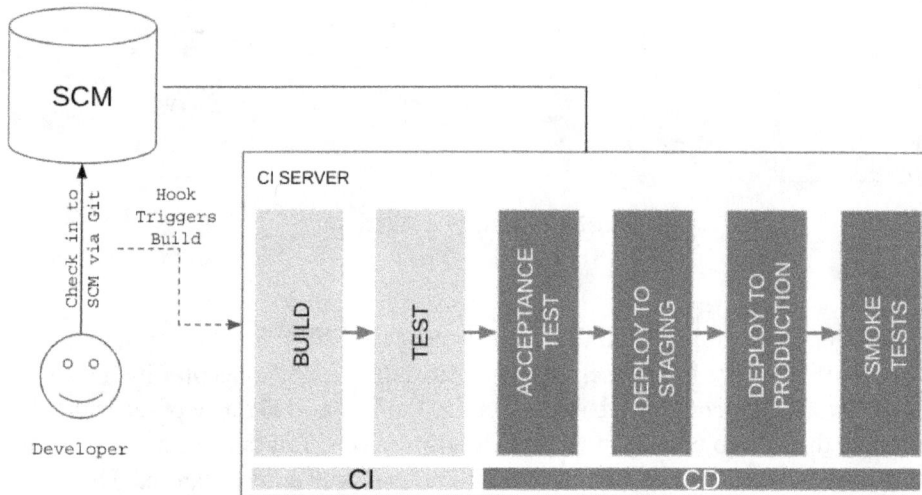

Figure 2.20 – CI/CD pipeline

Continuous integration and continuous delivery are advanced skills that you will gain over time. If you are looking for entry-level DevOps engineering roles, chances are they are not going to require you to have practical experience with CI/CD. However, it will be expected that you are able to discuss it and show an interest in it as it will likely be a large portion of your job. A good way to get started with CI/CD is to incorporate a pipeline into one of your code repositories. Have fun learning!

Cloud-native frameworks

Cloud-native is an approach to software development that leverages the capabilities of the cloud, both public and private. DevOps engineers will be involved in the use of cloud-native technologies in any career they choose, making it a very important and sought-after skill.

> **The Cloud Native Computing Foundation's (CNCF) Cloud-Native Definition:**
>
> Cloud-native technologies allow everyone to use immutable technologies with modern environments. Containers, service meshes, microservices, immutable infrastructure, and declarative APIs personify this approach and enable independent applications that are fault-tolerant and easy to manage. With automation, they enable engineers to make frequent changes with little disruption.

Cloud-native has several advantages, including faster development times and the ability to respond to customers more quickly.

Containers

A container is a lightweight purpose-built application that has been packaged with all the required dependencies for runtime so it can easily be run on any operating system in any environment with few changes needed. Multiple containers can run on the same machine while running as segmented processes in the user space. Containers take up less space than VMs, can handle more applications, and require fewer VMs and operating systems. The following is a diagram that compares the infrastructure needed for multiple VMs versus multiple containers:

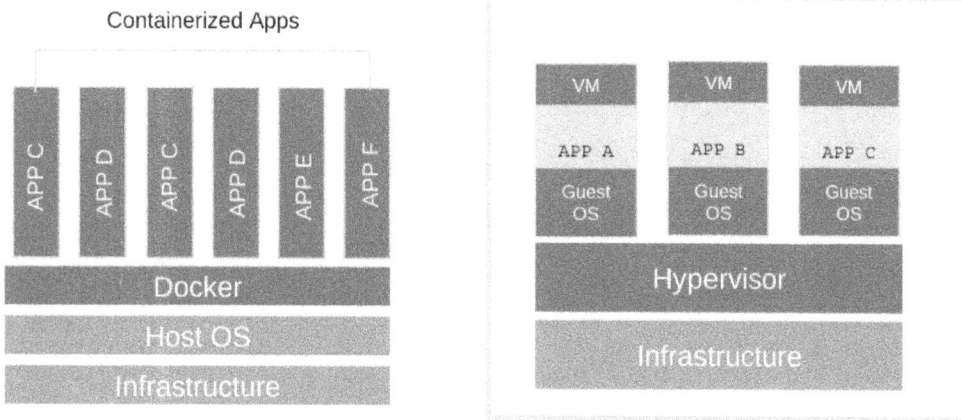

Figure 2.21 – Container and VM comparison

Docker exercise

In the following exercise, we will go through a basic Docker example, which can be done on your own computer.

The steps are as follows:

1. Follow this tutorial to install Docker on your machine: `https://docs.docker.com/get-docker/`.

2. Create a Dockerfile:

    ```
    touch Dockerfile
    ```

3. Add content to the Dockerfile using the `vi` editor:

```
 vi Dockerfile                        ×    +                          _  □  ×

FROM alpine:3.4
RUN apk update
RUN apk add vim
RUN apk add curl
~
~
~
~
~
~
~
~
-- INSERT --                                          4,17            All
```

Figure 2.22 – Dockerfile

After you have added the previous four lines to your file, make sure to save the file using `:w` and then exit `vi` using `:q`.

The following is the Docker image:

```
→ docker_exercise docker build --quiet --tag devopsbook:latest .
sha256:05dc32120cf7b2e63be5c9117c971e1f82d39ee3dabd4c47502076045d3b909e
→ docker_exercise docker images
REPOSITORY          TAG              IMAGE ID           CREATED
        SIZE
devopsbook          latest           05dc32120cf7       8 seconds ago
        32.2MB
alpine              3.4              b7c5ffe56db7       2 years ago
        4.81MB
→ docker_exercise
```

Figure 2.23 – Docker images

If you would like to see what exactly is happening during the build, you can omit the -- quiet command. After the image is built and tagged, you can see that the base alpine image along with the DevOps book image is available.

4. Run your container with the interactive terminal command:

```
→ docker_exercise docker run -it devopsbook:latest
/ # ls
bin       etc       lib       media   proc     run      srv     tmp      var
dev       home      linuxrc   mnt     root     sbin     sys     usr
/ # exit
→ docker_exercise
```

Figure 2.24 – interactive terminal command

The -it command runs the container with an interactive terminal, meaning a terminal session for the container will be opened, allowing you to interact with the container.

5. Stop and remove all containers and images from your machine:

```
→ docker_exercise docker stop $(docker ps -a -q)
9e5860d63800
→ docker_exercise docker rm $(docker ps -a -q)
9e5860d63800
→ docker_exercise
```

Figure 2.25 – Docker image removal

In this chapter, you learned about containers and the role they play in DevOps. If you followed along with the exercise, you will have created a Dockerfile, created a Docker image, and run the Docker image on your computer! Hopefully, this has made you interested in continuing to learn more about Docker as the rabbit hole goes deep!

Microservice architecture

Before looking at the desired architecture, we'll cover other dated architectures.

Monolithic

We will first cover the monolithic architecture, which shares a single code base and database. Because nothing is separate, everything must be released/deployed at the same time, which leads to long lead times between customer requests and them making it to production. I have worked at several companies and every one of them has had monolithic applications. There is a good chance you will run across this in your career.

Service-oriented

Service-oriented architecture (SOA) was a step in the right direction – it broke code down by services, which decreased the effort and time it took to get changes to production. Service-oriented architecture is prone to similar problems that a monolithic architecture has, such as interdependencies that require the entire application to be rebuilt even when a single service is checked in. Most of the companies I have worked with have had applications that use SOA.

Microservice

Microservice architecture has exploded in popularity due to large tech giants such as Amazon, Netflix, and Google publishing success stories about its use. The key differences between SOA and microservices are the communication protocols, storage, and size. Firstly, microservices use a language-agnostic protocol to communicate with the UI, resulting in a higher number of remote calls but also much higher fault tolerance. Secondly, each microservice has its own storage/database, which means each microservice can be designed with the right-fit database for its needs versus using the same database that is used for the entire application. Lastly, the size and lack of interdependence are what really separates a microservice from an SOA. A microservice can be deployed at any time and have no effect on other components. SOA shares a database and individual services still maintain some dependencies, which does not allow for the deployment of the service individually. Most companies are striving for a microservice architecture, which is why it is a key skill to have.

In the following diagram, you can see how monolithic, SOA, and microservice architecture compare:

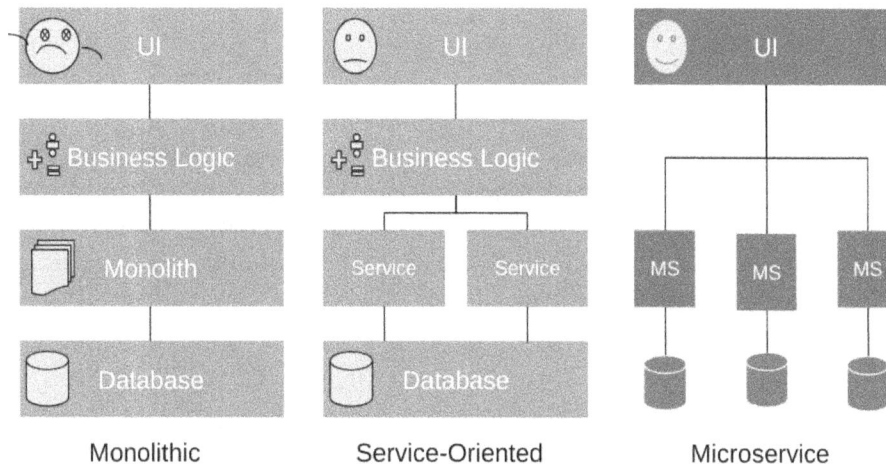

Figure 2.26 – Architecture comparison

Next, we will move on and discuss the importance of soft skills for DevOps engineers.

Soft skills

Soft skills are defined as *personal attributes that enable someone to interact effectively and harmoniously with other people*. The tides are changing, and DevOps engineers can no longer expect to succeed solely based on technical abilities. The following are a few of the most important soft skills for a DevOps engineer.

Empathy

With regard to DevOps, empathy is the ability to understand coworkers' and clients' points of view, or rather, the ability to view a situation from someone else's shoes. Approach your colleagues with a calm demeanor. This will lead to a much more pleasant work environment where new ideas flourish. If your idea differs from your colleague's or client's, start with positive feedback about their idea and work your way on to what you disagree with. Developing empathy with your coworkers ensures everyone's ideas are heard and issues that may be present can be resolved. Developing empathy with clients ensures all feedback is captured and a satisfactory end solution is reached.

Teamwork

Working in a team setting allows for multiple sets of eyes to view code at the same time. Working together ensures everyone remains on the same page and that a coherent product is delivered.

Adaptability

Tech is continuously changing and, as a DevOps engineer, you must prove you are good at changing gears, whether it be learning a new language or quickly shifting priorities. During an interview, you can discuss how you are learning a new programming language or how you have partnered with various departments while solutioning your last project. If you are unwilling to change, you will not succeed in DevOps.

Good communication

Good communication includes everything from in-person conversation to Slack messages. As a DevOps engineer, you will likely be working with team members who are fully remote, in different time zones, and coming from different cultures. So, you must be able to communicate effectively in each of these cases. Remember, people are busy, so pick the method that is going to be most efficient and effective.

Without strong soft skills, it will be difficult to land a job. DevOps is a team sport that requires you to collaborate with many different people in an environment that changes quickly. There is no room for drama or ego; everyone's opinion matters, and you need to respect that.

Beginner DevOps certifications

Like other industries, DevOps has seen an increase in the number of certifications that are available to practitioners. Certificates are a great way to showcase the knowledge you have, but they do not replace experience and are not required or mandatory to get a job as a DevOps engineer. DevOps certifications can help you stand apart from other candidates during the interview process. They also show your desire to continuously learn. When it comes to review time, you can use new certifications you have received since your last review as leverage for more merit. The following is a list of different certificates you can opt for:

AWS certifications

AWS offers a few entry-level certifications for DevOps engineers, starting with the AWS Cloud Practitioner certification, which requires about 6 months of hands-on experience with AWS. After you finish your AWS Cloud Practitioner exam, you can begin preparing for the AWS Associate Architect exam. The following are the certifications offered:

- AWS Cloud Practitioner (`https://aws.amazon.com/certification/certified-cloud-practitioner/`)
- AWS Associate Architect (`https://aws.amazon.com/certification/certified-solutions-architect-associate/`)

Google Cloud certifications

For Google, there is no generic beginner certification, but the Associate Cloud Engineer certification is a good introduction to GCP.

Associate Cloud Engineer (`https://cloud.google.com/certification/cloud-engineer`)

Azure certifications

For Azure, there is a fundamentals certification that covers a lot of the basics. There are several more, but we will cover more cloud certifications in the *Specialized competencies* section in the next chapter.

Fundamentals (`https://docs.microsoft.com/en-us/learn/certifications/exams/az-900`)

Other resources

For beginner content, there are a lot of courses from Udemy, EdX, and Coursera as well, depending on the specific areas you are interested in. Certifications that cover Docker, Terraform, and Kubernetes, as well as the advanced cloud specialties and professional certifications, will be covered in the following chapter.

Summary

In this chapter, you learned the basic skills required to succeed as an entry-level DevOps engineer. The skills included navigation of a text-only terminal such as Bash, automation using various scripting languages, and understanding Git and source code management. You also learned about the basics of infrastructure management tools such as Ansible and Terraform, as well as gaining an understanding of CI and CD and pipelines.

In the next chapter, you will dive deeper into the skills required for various DevOps specialty roles.

3
Specialized Skills for Advanced DevOps Practitioners

As DevOps engineers progress in their career-specific areas, they may stand out either as a result of their natural ability or skills or on account of their strong liking for the subject. This leads DevOps engineers down a specialized career path. The focus of this chapter will be on the skills required for entry into different DevOps specialties.

> **Mid–Senior Level Content**
>
> This chapter is full of useful information for anyone interested in the field of DevOps; however, it is geared for DevOps engineers who have been practicing for a minimum of 1-3 years. This chapter assumes that individuals already have the knowledge listed in the previous chapter.

In this chapter, the following specialized DevOps competencies will be covered:

- CI/CD pipeline DevOps engineer
- Infrastructure as code
- Cloud and application modernization

- Containers and container management
- Security
- Advanced DevOps certifications
- Competency matrix

CI/CD pipeline DevOps engineer

A CI/CD pipeline DevOps engineer is responsible for the end-to-end automation of the developer's code through to production. The strategy developed by the CI/CD engineer is a central focus for a company's DevOps roadmap. Revisiting the infinite DevOps loop from the previous chapter, you are reminded that all stages of DevOps are included in the CI/CD cycle, as seen in the following diagram:

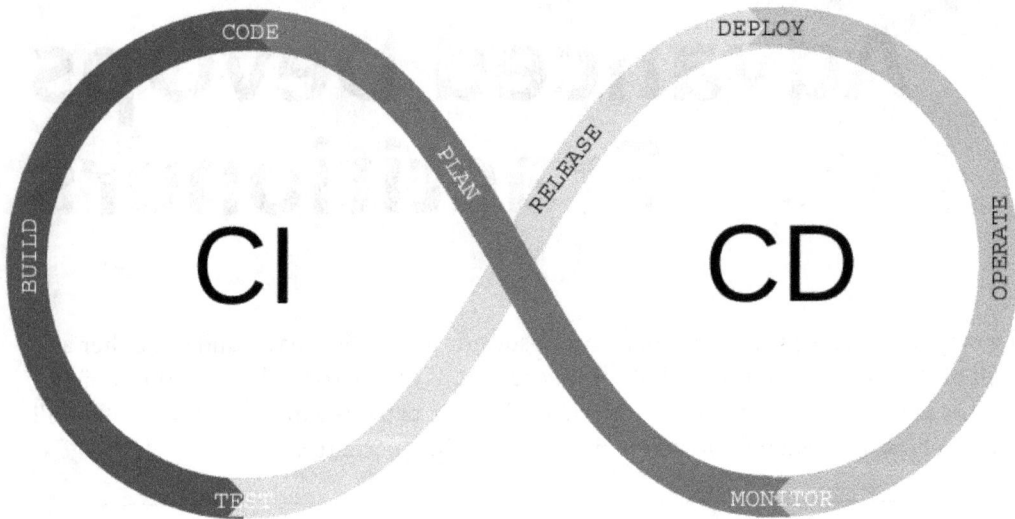

Figure 3.1 – Infinite DevOps cycle

One of the most in-demand skills for a CI/CD DevOps engineer is the ability to create, maintain, and promote the adoption of a shared pipeline library.

Maintaining a shared pipeline library

The CI/CD pipeline engineer will maintain the shared pipeline library and all the standards and practices that are associated with the CI tools for every stage of the DevOps life cycle.

> **CI Tool Management**
>
> Depending on the size of the company, responsibility for CI tool management may fall on the pipeline engineer; for this book, we will cover tool management in the *Cloud and application modernization* section.

To maintain the shared pipeline, DevOps engineers must have an expert-level understanding of the pipeline tools and pipeline architecture. Maintaining the shared pipeline library also involves managing and maintaining an inner-sourced project. As innovative ideas are proposed to be added to the pipeline, it is the role of the DevOps engineer to ensure it is implemented correctly.

> **Definition: Inner Source**
>
> Inner source is the use of open source software development best practices and the establishment of an open source-like culture within organizations for the development of its non-open source and/or proprietary software.

The following diagram is a graphical representation of how a shared pipeline library could work:

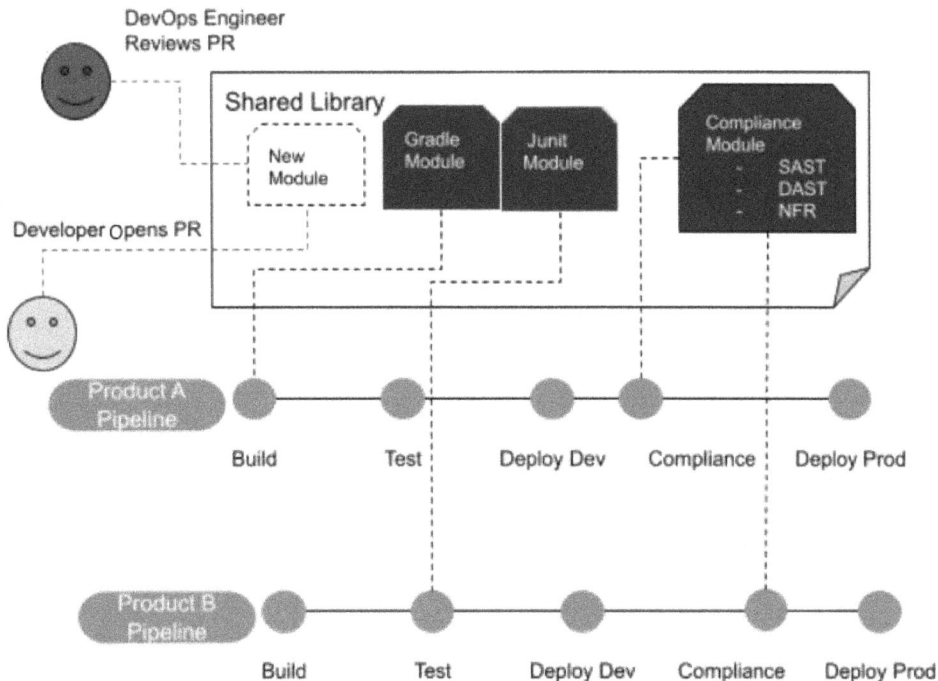

Figure 3.2 – Shared pipeline library

A lot is going on in the preceding diagram; first, we have a shared pipeline library that has different component modules in it that can be used by various products. We have a developer who developed some new functionality that they thought would be useful for the rest of the company, so they opened a pull request, which needs to be reviewed by the CI/CD DevOps engineer before it officially becomes part of the library. Lastly, you can see two distinct products using components of the shared library in their own pipelines.

Ownership of integrations with the pipeline

A DevOps pipeline engineer needs to have a vast knowledge of tooling at all stages of the pipeline. The following diagram shows many of the tools available at various stages of the pipeline:

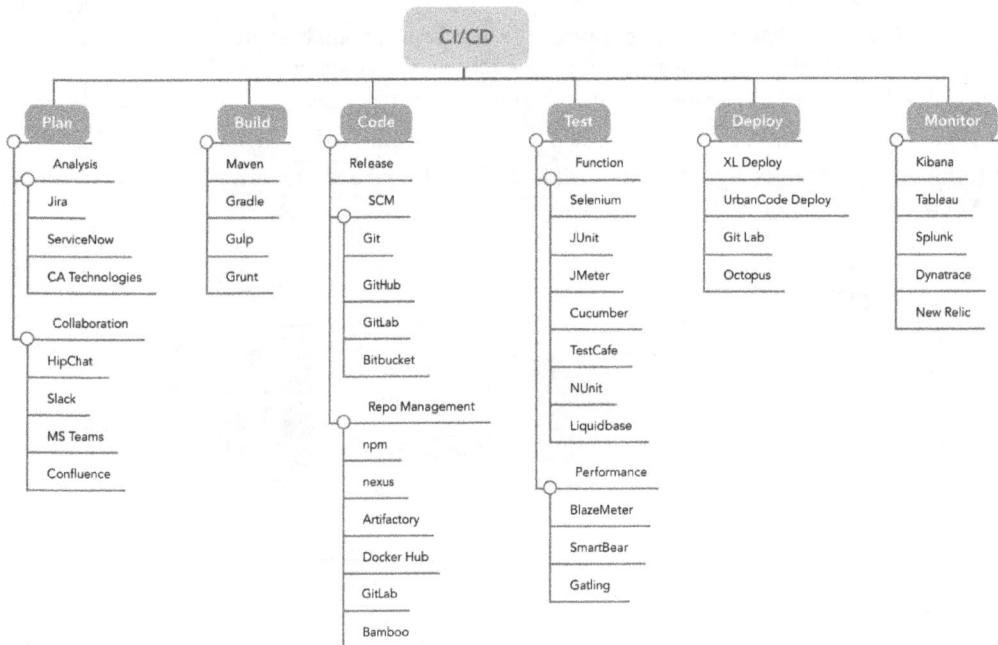

Figure 3.3 – CI/CD tools

The preceding diagram is meant to demonstrate the vast number of tools available at each stage of the CI pipeline. A valuable resource that allows you to view and examine each tool is digital.ai's periodic table of DevOps tools (`https://digital.ai/periodic-table-of-devops-tools`).

The CI/CD DevOps engineer is responsible for ensuring that the various tools can be integrated into the company's pipeline. Modules for connecting to the tool's API must be developed and then the solution must be presented to a security review to ensure that it meets companies' standards. Once it is part of the pipeline, the CI/CD engineer is responsible for ongoing support of the various integrations as the tool's version and API changes.

In this section, you learned that a CI/CD DevOps engineer must be knowledgeable in all phases of the CI/CD life cycle and have an understanding of the various tools used in each of the stages.

In the next section, we will cover the skills required to be successful as a DevOps infrastructure engineer.

Infrastructure as code

A DevOps infrastructure engineer is responsible for provisioning, managing, and maintaining the infrastructure used within a company's various applications. Oftentimes, referred to as the **Infrastructure as Code (IaC)** engineer, the IaC engineer works closely with the architects and must have good relationships with many diverse groups. In this section, we will cover the skills required to be successful as a DevOps IaC engineer as well as the tools you will need to have a deep understanding of the role.

Network infrastructure design

DevOps IaC engineers partner closely with architects and, in some cases, play the role of the architect for many projects.

> **Network Infrastructure Design: Definition**
>
> Network infrastructure design is a process comprising network synthesis, topological design, and network realization, aimed at ensuring that a network or service meets the needs of the operator and subscriber.

Storage management

DevOps IaC engineers need to be knowledgeable on the broad topic of storage management and optimization. Storage management covers volume migration, process automation, disaster and recovery, replication, auto-provisioning, snapshot and mirroring, storage virtualization, and compression.

Containerization (Docker and Kubernetes)

DevOps engineers need to be familiar with containers; however, DevOps engineers specializing in infrastructure management will need to be comfortable with the orchestration and management of large Docker and Kubernetes clusters. Tools such as Terraform and Ansible can be used to help with this.

The topic of containerization will be covered in much more detail in the *Containers and container management* section of this chapter.

Site Reliability Engineering

Site Reliability Engineering (**SRE**) focuses on system availability and reliability and was coined by Ben Treynor of Google in 2003. The goal of SRE is the same as DevOps – to bridge the gap between development and operations. If you work for a large organization, SRE and DevOps will be separate entities on account of their goal and focus; however, the skills required for both SRE and a DevOps engineer focused on infrastructure will be remarkably similar. SRE has a focus on keeping systems running and available while DevOps aims to reduce the time to market and allow for rapid changes. In the following diagram, the items that should account for the most time are the foundation of the SRE hierarchy:

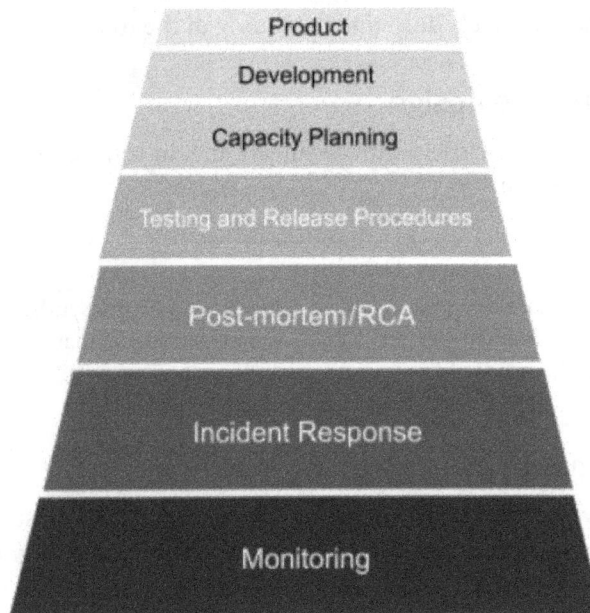

Figure 3.4 – SRE hierarchy

The preceding diagram illustrates how the foundation of SRE is a proactive, monitoring-based approach, followed by quick and thorough incident response and detailed *no-blame* postmortems. The idea behind SRE is to find and remediate problems quickly, followed by changes to avoid the same issue in the future.

In this section, we scratched the surface of a deep and interesting subject. If you are interested in learning more about SRE, you should take a look at the book on SRE published by Google: `https://sre.google/sre-book/table-of-contents/`. In the last section of this chapter, we will list several certifications that will apply to a role in SRE.

In this section, you learned what is required to be a DevOps engineer specializing in infrastructure. You need skills in network design, storage management, SRE, and containerization.

In the next section, we will cover the skills required by a DevOps engineer who specializes in cloud and application modernization.

Cloud and application modernization

In this section, you will learn about cloud and application modernization, and the special role that exists for DevOps engineers in this space. First, we will cover the advanced cloud knowledge required, followed by cloud modernization techniques.

DevOps leaders need to adjust to a new demand for updated services while maintaining, operating, and improving their existing application portfolios. This new demand comes from modern technology, which is being introduced with increasing frequency. There are many application modernization approaches (including rehost, replatform, and replace) with different purposes, effects, values, costs, risks, and impacts.

Advanced cloud skills

To be a DevOps cloud engineer, you must have strong scripting skills in addition to an in-depth understanding of cloud platforms. You need to understand how different cloud products function and work together. The following diagram contains something that comes in handy if you are working in an area that has a multi-cloud environment:

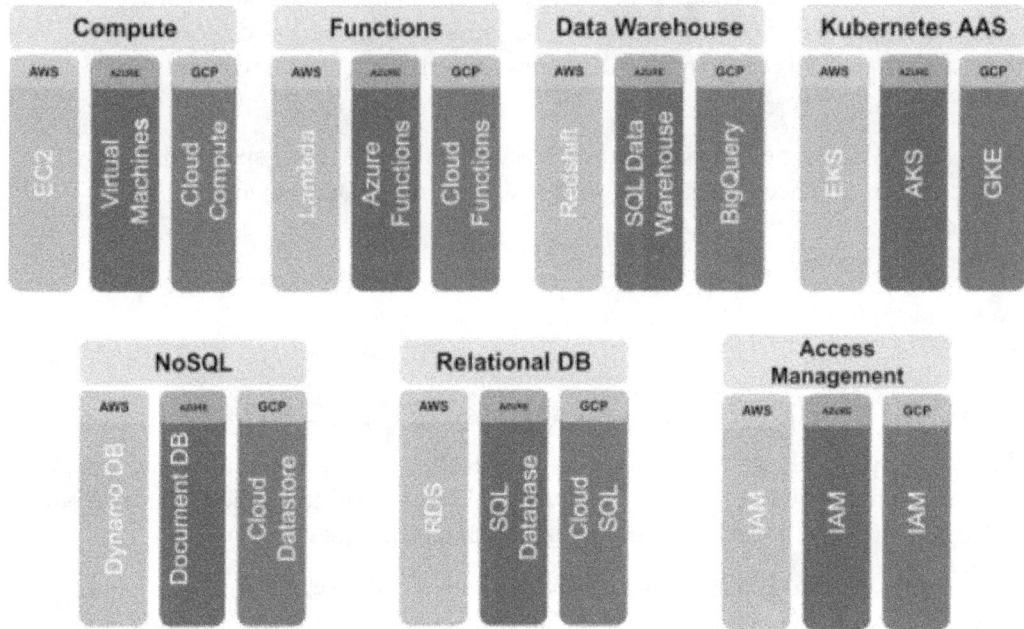

Compute				Functions				Data Warehouse				Kubernetes AAS		
AWS	AZURE	GCP		AWS	AZURE	GCP		AWS	AZURE	GCP		AWS	AZURE	GCP
EC2	Virtual Machines	Cloud Compute		Lambda	Azure Functions	Cloud Functions		Redshift	SQL Data Warehouse	BigQuery		EKS	AKS	GKE

NoSQL				Relational DB				Access Management		
AWS	AZURE	GCP		AWS	AZURE	GCP		AWS	AZURE	GCP
Dynamo DB	Document DB	Cloud Datastore		RDS	SQL Database	Cloud SQL		IAM	IAM	IAM

Figure 3.5 – Cloud provider comparison

I am very comfortable with Amazon terminology; when I needed to work in GCP, it was handy to have a cheat sheet nearby. The methodology is the same; the terminology is vastly different across cloud providers.

Another thing that DevOps cloud engineers need to be capable of is provisioning and deploying to cloud resources; this is another thing that varies greatly between different cloud providers. Understanding the cloud provider CLI is also very important as a DevOps engineer. You will rarely interact with resources via the **graphical user interface (GUI)**; instead, you will use the CLI tool or API to make calls to various cloud offerings. The following links relate to the documentation pertaining to the various cloud CLI tools:

- GCP: `https://cloud.google.com/sdk`

- AWS: `https://aws.amazon.com/cli/`

- Azure: `https://docs.microsoft.com/en-us/cli/azure/install-azure-cli`

In the next section, we will go through another aspect of cloud specialty – application modernization. The topic is dense and there is no effective way to get good at it without practicing it; however, Gardner provides some useful resources for individuals looking to learn more.

Application modernization

As a DevOps engineer who specializes in cloud and application modernization, evaluating the driving force behind the need for modernization is crucial. The driving force can be broken down into the demand and supply sides:

- **Supply side**:

  - **Fit**: Lacking the ability to implement new requirements

  - **Value**: Lacking in terms of the value, quality of support, and information it provides

  - **Agility**: Lacking the ability to make changes quickly with an acceptable level of risk

- **Demand side**:

  - **Cost**: The cost of ownership is high in relation to the value it provides.

  - **Complexity**: Complexity causes maintainability and increases risk when making changes.

  - **Risk**: Security, compliance, supportability, or scalability risk.

After a DevOps engineer has gone through and evaluated the system and determined the issue, the causes must be identified.

Choosing a modernization approach

The cause for modernization can be broken down into three main categories: functionality, technology, and architecture. Once the cause is determined, the DevOps engineer must decide on the best approach for modernization. The following is a diagram showing the different modernization approaches with respect to their effort and complexity:

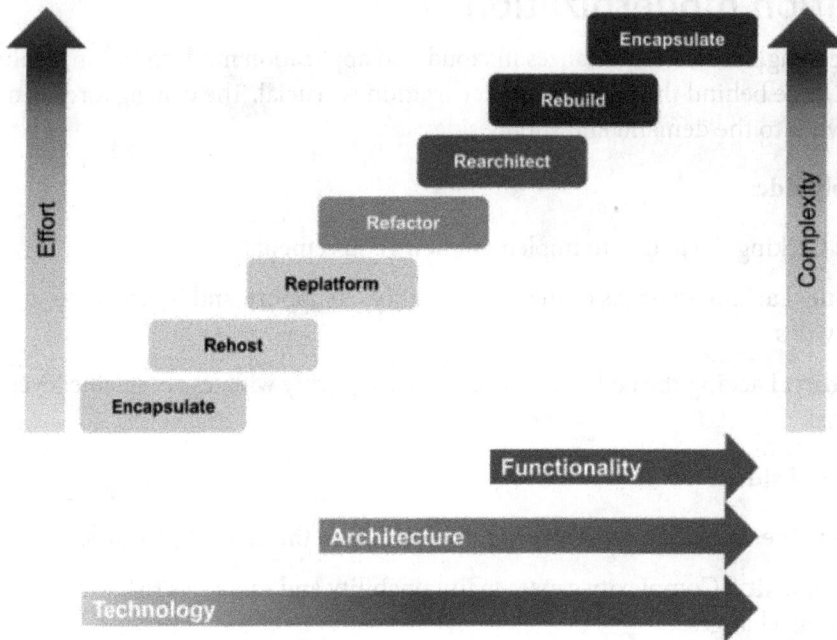

Figure 3.6 – Gartner's modernization approaches' ability to remediate the cause

As a DevOps engineer who specializes in cloud and application modernization, you will be required to have an in-depth understanding of different modernization approaches. Gartner published an article, `https://www.gartner.com/doc/reprints?id=1-25RJ3RG2&ct=210408&st=sb`, which covers in detail each of the aforementioned approaches.

In this section, we covered the skills required for a DevOps engineer specializing in application modernization as well as the cloud. A DevOps engineer specializing in the cloud must be extremely comfortable provisioning and deploying to cloud applications. DevOps engineers will need to have a strong understanding of the cloud API and CLI functionality as well as understanding how cloud offerings compare to other cloud providers. Additionally, DevOps engineers specializing in the cloud may be required to aid in application modernization projects.

In the next section, I will discuss the skills required to be successful as a DevOps engineer specializing in containers.

Containers and container management

Container management, orchestration, and maintenance are skills all DevOps engineers must be competent in; however, with cloud-native Kubernetes and other cloud orchestration tools, it has quickly become a specialty field. In this chapter, we will cover the skills required to succeed as a DevOps engineer specializing in container management.

> **What It Takes to Be a Container Specialist**
>
> The ability to pioneer, adopt, and own new container methodologies is the one skill that separates a generalist from a container management specialist.

Container management software

DevOps engineers who specialize in containerization must be highly skilled in container management software. The most widely used software for container management include Docker and Kubernetes. The purpose of container management software is fivefold; automation, monitoring, security, scaling, and deployment for containers. We will cover the automation aspect first:

- **Container automation**: The automation of container orchestration encompasses and covers monitoring, security, scaling, and deployment at an elevated level.

- **Monitoring**: Containers require monitoring, such as liveliness probes, which are tied to monitoring dashboards and alarms that can be used to spot and remediate problems sooner. There are several tools that can be used for monitoring, including Prometheus, Dynatrace, and Datadog.

- **Security**: Container security is the most critical concern for a DevOps engineer as the entire application is exposed if the container is compromised. To ensure that your container ecosystem is secure, first ensure that the base images being used are secure. Make sure the container images are signed from trusted sources, that the runtime operating system layer is up to date, and ensure that you have a patching strategy in place to update the image. Next, you need to ensure that your images are stored in a secure location with policy-based authentication to reduce the chance of human errors being introduced into the containers. Integrating security testing and scans and automating the deployment process is another way to mitigate risk. Tools such as Twistlock can help with this. Scanning the container allows for containers to be rebuilt and redeployed instead of trying to patch running containers.

- **Scaling**: Containers, by design, are immutable, which means that code cannot be changed following deployment. A benefit of containers is that with proper configuration and monitoring, scaling the number of instances of a certain component up or down is an easy task. The hard part is determining the limits for when such scaling should occur.

This section described what is required to be a DevOps engineer who specializes in containers. In the conclusion to this section, I leave you with a few hands-on tutorials I used while learning Kubernetes:

- Kubernetes the hard way: `https://github.com/kelseyhightower/kubernetes-the-hard-way`

- Kubernetes training course: `https://www.udemy.com/course/certified-kubernetes-application-developer/`

In the next section, we will cover what is required to succeed in the field of DevSecOps, or, in other words, a DevOps engineer specializing in security.

Security

A DevOps engineer who specializes in security is often referred to as a DevSecOps engineer. DevSecOps engineers have a deep understanding of the CI/CD process as well.

> **Security Is Everyone's Responsibility**
>
> Security is everyone's job. Anyone who has any stake in delivering the software has a role in ensuring the application's security.

The job of a DevOps security engineer is to ensure that security is built in and included from the onset of a project.

DevOps engineers specializing in security have responsibilities that are broken down into two areas: CI/CD processes and environment and data. We will first look at CI/CD process security and the skills required to implement it.

CI/CD process security

Let's revisit the pipeline discussed earlier; numbers have been added to the following diagram to correlate to the following security items:

- **Container scanning** (*1*): Container scanning should be added to the process of bringing new containers into your registry. Tools such as Twistlock are commonly used. The most popular container scanning tool is Twistlock; however, there are many others as well and these can be seen here: `https://techbeacon.com/security/17-open-source-container-security-tools`.

- **Security testing** (*2*): This includes running **security static analysis testing** (**SAST**) tools as part of builds, as well as scanning any pre-built container images for known security vulnerabilities as they are pulled into the build pipeline. In the following example, the libraries are part of the compliance module. Many tools exist, both open source and paid, when it comes to SAST tools.

- **Security acceptance testing** (*3*): This type of testing is known as **dynamic application security testing** (**DAST**). The purpose is to test for security vulnerabilities dynamically while the application is running. A comprehensive tools list in this category can be found here: `https://owasp.org/www-community/Vulnerability_Scanning_Tools`.

- **Patching and security updates** (*4*): Patching and security updates should have a pipeline so that they can run automatically and at a scheduled cadence. The tools that can be used vary greatly. In my experience, the best patching tool is sometimes a simple script executed in a pipeline. Do your own research and form your own opinions.

- **Configuration management** (5): The purpose of this is to ensure that all infrastructure follows company security and compliance policies. Oftentimes, this is run when infrastructure is provisioned or after config changes have been applied.

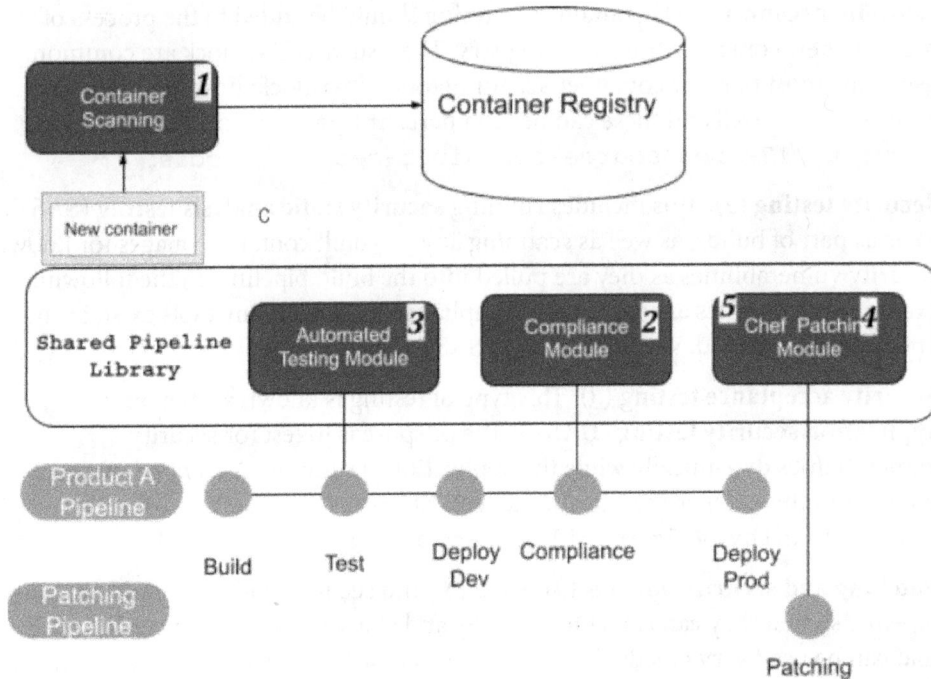

Figure 3.7 – CI/CD security

Integrating security into CI/CD pipelines is tricky as it requires domain knowledge as well as tool-specific knowledge and an understanding of industry security best practices. One article that helped me better understand compliance pipelines can be found at the following link: `https://itrevolution.com/book/devops-automated-governance-reference-architecture/`. In the next section, we will cover the areas DevSecOps engineers need to focus on to ensure environment and data security.

Environment and data security

The various environments and associated data are a company's most precious and vulnerable assets if not protected correctly. It is the DevOps engineer's job to ensure the best practices are implemented in an automated and scalable fashion. The following are a few concepts that will help drive further research on the topic:

- **Principle of least trust and zero trust**: Processes and applications are granted the minimum access they require to operate properly. It sounds simple but it is quite a large undertaking as each account needs to be audited to ensure it has correct access. The principle of least privilege is a core part of the zero-trust model; however, the zero-trust model is more comprehensive and is more stringent. Also, zero trust is more complicated to implement and maintain as many more access policies are needed. The following diagram shows the three principles of zero trust; verify explicitly, use least privilege access, and assume breach:

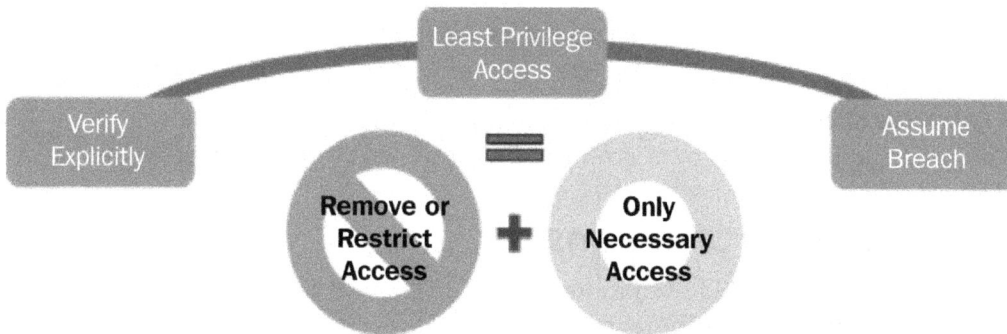

Figure 3.8 – Zero-trust principles

- **Principle of isolation**: The principle of isolation applies to containers running in a microservices architecture. The principle of isolation has several components, including isolation of state, isolation of space, isolation of time, and isolation of failure.

- **Data encryption**: A container orchestration platform with integrated security features helps minimize the chance of unauthorized access.

- **Secure API gateways**: Secure APIs increase authorization and routing visibility. By reducing exposed APIs, organizations can reduce attack surfaces.

Advanced DevOps certifications

As you progress as a DevOps engineer, the number of certifications available to you increases as well. Certifications are rarely required to get a job in a specialized DevOps role, but they do offer a quick picture of the skills and expertise you possess as well as showing your dedication to your industry.

AWS certifications

AWS offers several advanced certifications based on what career track you are on. Most advanced AWS certifications require a minimum requirement of possessing the AWS Cloud Practitioner certification. The following is a list of the certifications offered:

- **AWS Certified Solutions Architect – Professional** (`https://aws.amazon.com/certification/certified-solutions-architect-professional/`)

- **AWS Certified DevOps Engineer – Professional** (`https://aws.amazon.com/certification/certified-devops-engineer-professional/`)

- **AWS Certified Security – Specialty** (`https://aws.amazon.com/certification/certified-devops-engineer-professional/`)

Google Cloud certifications

Google offers several advanced certifications related to DevOps depending on what specialty you are interested in. The following is a list of the certifications offered:

- **Professional Cloud Architect** (`https://cloud.google.com/certification/cloud-architect`)

- **Professional Cloud DevOps Engineer** (`https://cloud.google.com/certification/cloud-devops-engineer`)

- **Professional Cloud Security Engineer** (`https://cloud.google.com/certification/cloud-security-engineer`)

- **Professional Cloud Network Engineer** (`https://cloud.google.com/certification/cloud-network-engineer`)

Azure certifications

Azure offers numerous advanced certifications, including one in DevOps and one in security technologies. The path to DevOps certifications is AZ-900, AZ-104, and then AZ-400. The path for an architect is AZ-90, AZ-300, and then AZ-304. The path for security is AZ-900 followed by AZ-500. The following is a list of the certifications offered:

- **Microsoft Azure Administrator** (`https://docs.microsoft.com/en-us/learn/certifications/exams/az-104`)

- **Microsoft Azure Architect Technologies** (`https://docs.microsoft.com/en-us/learn/certifications/exams/az-303`)

- **Designing and Implementing Microsoft DevOps Solutions** (`https://docs.microsoft.com/en-us/learn/certifications/exams/az-400`)

- **Microsoft Azure Architect Design** (`https://docs.microsoft.com/en-us/learn/certifications/exams/az-304`)

- **Microsoft Azure Security Technologies** (`https://docs.microsoft.com/en-us/learn/certifications/exams/az-500`)

Kubernetes certifications

Kubernetes is one of the largest open source projects and what employers are looking for as regards a top skill when hiring DevOps engineers. The following is the certification offered:

- **Certified Kubernetes Administrator (CKA)** (`https://www.cncf.io/certification/cka/`)

In the following diagram, you can see the certification paths for the three major cloud providers – AWS, Azure, and GCP:

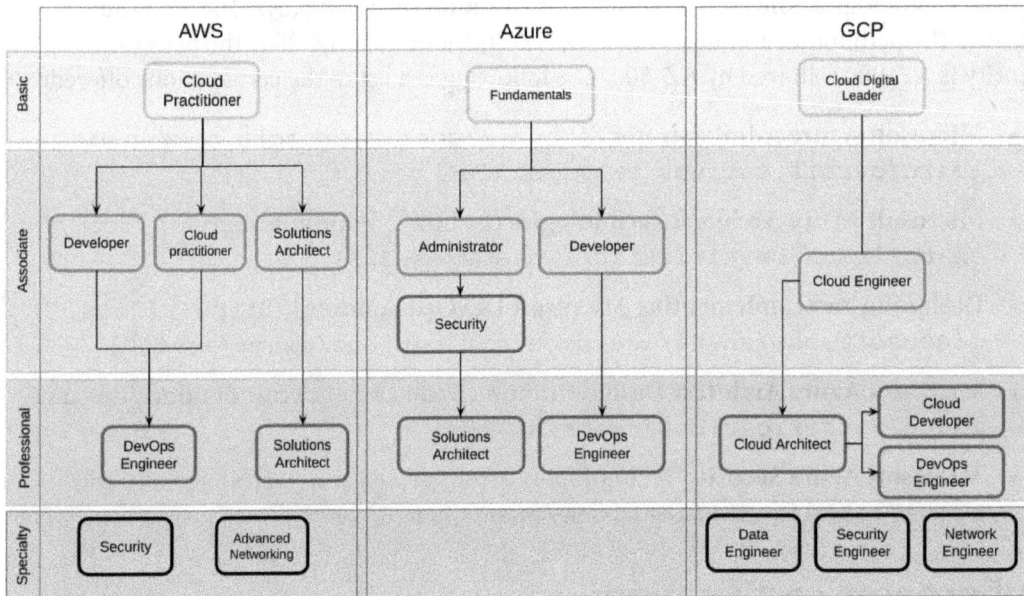

Figure 3.9 – Cloud certification path

In the next section, we will look at a skills matrix for DevOps engineers.

Competency matrix

A competency matrix is a visual diagram or chart that shows the skills and education required for separate roles within a company. Before we dive into competency matrix definitions, a few things need to be understood.

First, not all organizations use a competency matrix to hire and promote employees, and second, each organization that does use it will have different requirements.

The reason it was decided to have a section on competency is that every company has differing competencies tied to distinct levels. What this means is that a senior DevOps engineer at company X may map to a DevOps engineer at Google. This caused me anxiety and frustration early in my career. I hope this section can give you enough clarity on the correlation between level, competency, and pay.

The purpose of this chapter is to introduce you to the concept of a competency matrix as well as adding another tool to track your progress as your skills increase. We will start by breaking down the competency matrix.

Matrix breakdown

A competency matrix is broken down into three main sections: **skills**, **levels**, and **competency**.

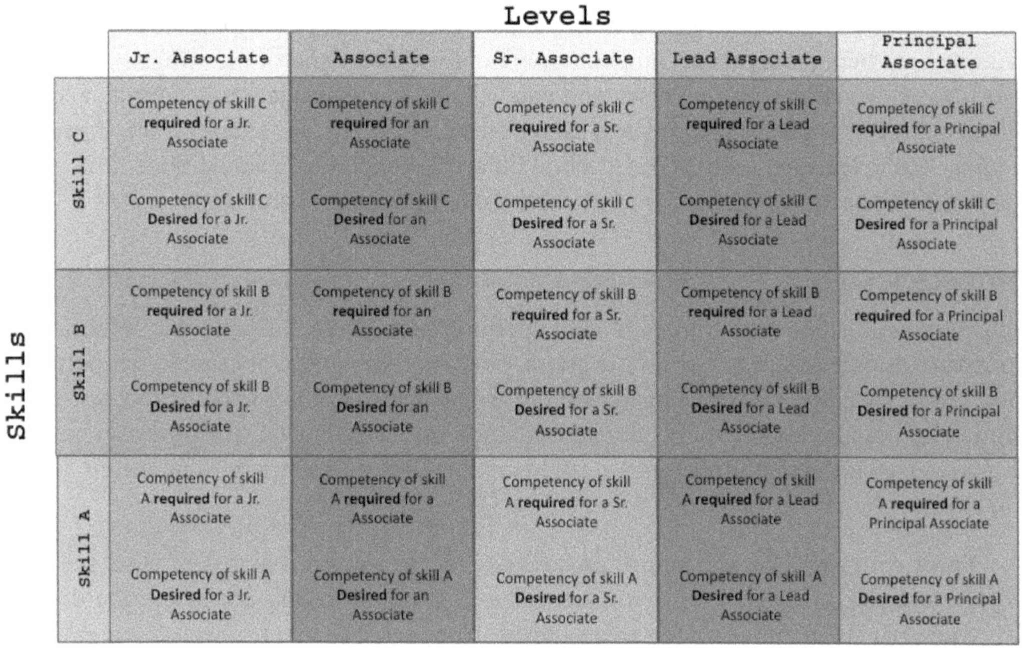

		Levels			
	Jr. Associate	**Associate**	**Sr. Associate**	**Lead Associate**	**Principal Associate**
Skill C	Competency of skill C **required** for a Jr. Associate Competency of skill C **Desired** for a Jr. Associate	Competency of skill C **required** for an Associate Competency of skill C **Desired** for an Associate	Competency of skill C **required** for a Sr. Associate Competency of skill C **Desired** for a Sr. Associate	Competency of skill C **required** for a Lead Associate Competency of skill C **Desired** for a Lead Associate	Competency of skill C **required** for a Principal Associate Competency of skill C **Desired** for a Principal Associate
Skill B	Competency of skill B **required** for a Jr. Associate Competency of skill B **Desired** for a Jr. Associate	Competency of skill B **required** for an Associate Competency of skill B **Desired** for an Associate	Competency of skill B **required** for a Sr. Associate Competency of skill B **Desired** for a Sr. Associate	Competency of skill B **required** for a Lead Associate Competency of skill B **Desired** for a Lead Associate	Competency of skill B **required** for a Principal Associate Competency of skill B **Desired** for a Principal Associate
Skill A	Competency of skill A **required** for a Jr. Associate Competency of skill A **Desired** for a Jr. Associate	Competency of skill A **required** for a Associate Competency of skill A **Desired** for an Associate	Competency of skill A **required** for a Sr. Associate Competency of skill A **Desired** for a Sr. Associate	Competency of skill A **required** for a Lead Associate Competency of skill A **Desired** for a Lead Associate	Competency of skill A **required** for a Principal Associate Competency of skill A **Desired** for a Principal Associate

(Skills — vertical axis label)

Figure 3.10 – Competency matrix

First, we will cover the skills section.

Skills

The skills section is usually on the vertical (y) axis of the chart. The skills can be as granular or as broad as a company sees fit. The skills represented on the diagram will be specific for the department you are part of, so it is best practice to stay updated and current on your department's competency matrix if it exists. Next, we will discuss the horizontal axis of the matrix.

Levels

The level is located on the horizontal (x) axis of the chart. Levels differ from one company to the next. I currently work for an organization where all individual contributors are mapped to different associate levels. If you were to work at Google, your role is mapped to L1 through L11 where an engineer is an L3, and a senior Google fellow is an L11. Levels without being mapped to competency are nothing more than a meaningless term. Unfortunately, levels are not universal, and you will need to investigate how your current skills map over if you decide to pursue a career outside your current organization. If you are interested in learning more about levels at different companies, a valuable resource is levels.fyi: `https://www.levels.fyi/`. Next, we will discuss competency.

Competency

Competency is defined as the ability to do something successfully. Competency is found at the crossroads of levels and skills on the graph. Each level has different criteria for success for each skill, meaning success is relative to the level you are at. It is becoming increasingly common for companies to adopt competency bands within each level. The purpose of this is to provide entry competency, desired competency, and sometimes premium competency:

- **Entry** competency is the minimum competency for a given skill to be considered for a role at the given level. If you are in the minimum category across the board, your chances of landing a job are not great. However, if there are only a few areas where you are in the minimum range, they can be a talking point when you are being interviewed of things you are working to improve on.

- **Desired** competency is the competency of about 50% of individuals in the level or, more simply, the competency the hiring team is hoping to find. The ideal situation for a hiring manager is that they find a candidate with desired qualifications across the board.

- **Premium** competency is the point where there begins to be a lot of creep between the current level and the next level up. If you are in the premium band for several skill areas, you are ready to be promoted to the next level.

The following diagram shows that competency overlap exists between levels. This feature is built to allow fluid movement between levels:

Levels

		Jr. Associate	Associate	Sr. Associate	Lead Associate	Principal Associate
Development	premium	Single task e2e	Understands entire feature	Takes ownership of a feature	Leads team working on feature that impacts entire product	Owns entire product
	desired	Single task with help from others	Single task e2e + mentors jr. associate	Understands entire product	Leads a team working on a feature	Leads multiple feature teams
	entry	Observes other associates	Single task e2e	Understands entire feature	Takes ownership of a feature	Leads team working on feature that impacts entire product

Figure 3.11 – Competency matrix levels

Your pay range is bound by your level, but within a level, skill competency determines where you land within the range. This is discussed in detail in the following section.

Compensation in relation to level and competency

Compensation is bound by pay bands that are tied to the various levels within the company. The pay bands for distinct levels overlap; this means that a senior associate could potentially earn a higher base salary than a lead associate. The following diagram does an excellent job graphically of depicting this:

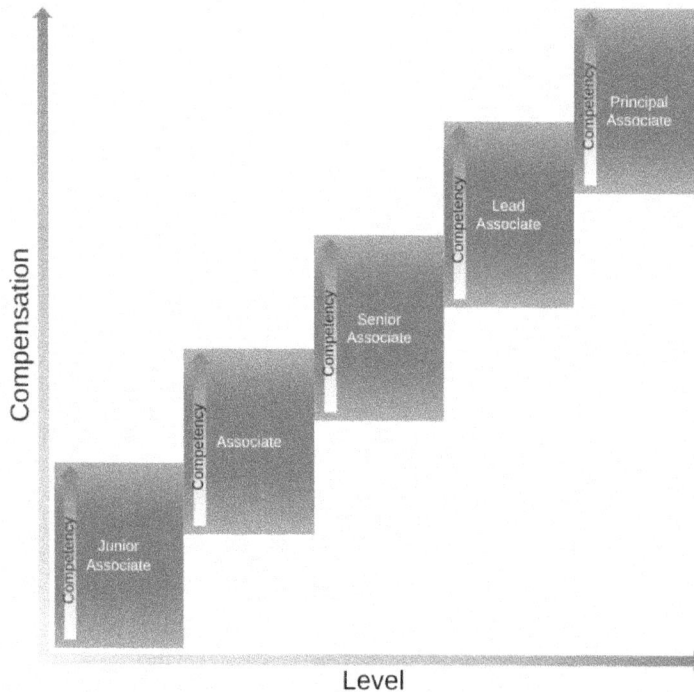

Figure 3.12 – Compensation – Level graph

In an effort to normalize pay within each level, managers are often told to target a specific percentile within each pay band. This is useful information when looking at job postings that have pay ranges.

Summary

In this chapter, you learned the skills required for DevOps specialist roles, including as a CI/CD pipeline specialist, infrastructure specialist, cloud and application modernization specialist, container specialist, and security specialist. The main takeaway from this chapter is that DevOps specialties take practice and dedication to hone your skills. We also discussed the competency matrix and how it translates to different pay grades.

In the next chapter, we will cover how to rebrand yourself online.

Section 2: The Application Process

From updating your résumé and online profiles through to follow-up emails after submitting your résumé to a recruiter, this section of the book will aid in every facet of applying for DevOps jobs.

This section comprises the following chapters:

- *Chapter 4, Rebranding Yourself*
- *Chapter 5, Building Your Network*
- *Chapter 6, Mentorship*
- *Chapter 7, Working with Recruiters*

4
Rebranding Yourself

You are continuously learning, growing, and becoming a better version of yourself. Those who see you on a frequent basis know exactly who you are and what skills you possess; however, to the rest of the world you are defined by your social profiles. In this chapter, we will guide you through how to ensure your social profiles, resume, and personal web pages match who you have become and what you want to accomplish. We will start by refreshing your social profiles, then work on your other online sites, such as GitLab and GitHub, and end with ensuring your resume matches your social profiles.

We will cover the following main topics in this chapter:

- Ways of improving your LinkedIn profile
- Updating your resume to match the career you are after
- Updating or creating your personal web page
- Leveraging Twitter and other social profiles

Ways of improving your LinkedIn profile

Your social profiles are seen by everyone, and recruiters are continuously looking at them. An updated and maintained **LinkedIn** profile is the easiest way to get noticed by recruiters and hiring managers. In this section, we'll cover all the changes needed to ensure your LinkedIn profile represents you best and is seen by more people.

Updating your headline

Your headline is highly visible to others on LinkedIn. By default, it displays your job title and where you work. Be creative, use it to show off your top skills, advertise your job search, or use a unique tagline to ensure your profile stands out from everyone else.

If you are an experienced DevOps engineer looking for a career change, you could try the following headline:

DevOps | Cloud | Containers | Open to Remote Opportunities

If you are not looking for a new career but want to stand out, try the following headline:

Experienced Cloud Engineer | AWS | GCP | AZURE

The following are two fictitious individuals who have bland and boring headlines on the left with bold and refreshing headlines on the right:

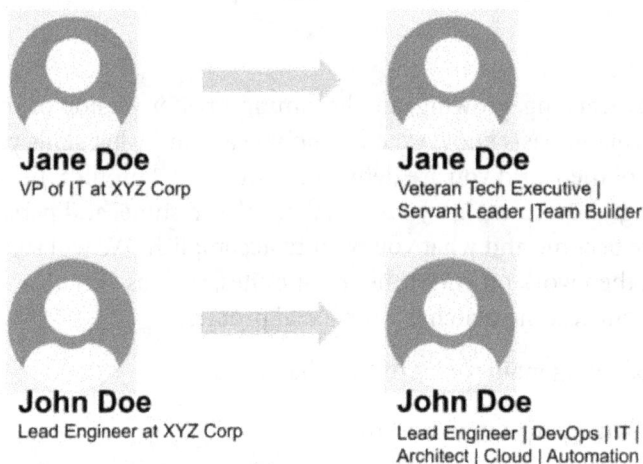

Jane Doe
VP of IT at XYZ Corp

Jane Doe
Veteran Tech Executive |
Servant Leader |Team Builder

John Doe
Lead Engineer at XYZ Corp

John Doe
Lead Engineer | DevOps | IT |
Architect | Cloud | Automation

Figure 4.1 – LinkedIn headline refresh

Your headline can be up to 220 characters and should describe what you are skilled at and why people should be interested in you. Having keywords, such as **DevOps**, **Cloud**, and **AWS**, will increase the number of search results you appear in on LinkedIn.

In the next section, we will discuss why recommendations are important and how to ask for them.

Recommendations

LinkedIn can request recommendations from colleagues and managers you have worked with, and then display them on your profile. This allows individuals who visit your profile to see recommendations from individuals who have worked with you before they even interview you. Here are a number of tips when asking for recommendations:

- Recommendations should come from people you have a good relationship with, preferably, someone you can ask to recommend you prior to sending the request on LinkedIn.

- Recommendations should come from people you respect and whose work is respected in the industry you are in.

- If you are a more senior DevOps professional, recommendations should come from individuals you have mentored, other senior DevOp peers, or a director or VP, if you have a close working relationship.

- Do not request a recommendation from someone you have not spoken with for an extended period or someone you parted with on bad terms.

- Recommendations from professors you are close with are immensely powerful.

- Do not request a recommendation from another junior DevOps engineer or individual who is also looking for their first role in the profession. Instead, ask your mentor (if you have one) for a recommendation. The importance of mentorship will be covered in *Chapter 6, Mentorship*.

If you receive a recommendation and are not satisfied with how it looks, it is OK to ask for it to be revised. Make sure to provide specific feedback and the reasons for requesting the change. A poorly written recommendation can do more harm than good. The following is an example of a poorly written recommendation:

john is good engineer at all tasks given to him.

Jane Doe
Veteran Tech Executive | Servant Leader |
Team Builder

Figure 4.2 – An incomplete and vague recommendation

The problem with the preceding recommendation is the poor grammar and lack of detail; what tasks is John good at, and why? If I saw this recommendation, I would be very curious; it came from a highly credible person but lacks any meaningful details. The following is the same recommendation with better grammar and more detail:

John is the quintessential team player. He is always among the first to volunteer, and will help support his teammates in whatever way he can. As a skilled engineer who understands the importance of technical practices such as Test-Driven Development, Continuous Integration, and Continuous Delivery; Brett understands how to build quality in and by doing so has saved parts unlimited 1000's of hours of rework .
See less

Jane Doe
Veteran Tech Executive | Servant Leader | Team Builder

Figure 4.3 – A well-written recommendation

To summarize this section, recommendations are important and can help you land a job when they come from a respected individual and are written well. In the next section, we will discuss the sections that you should include in your LinkedIn profile but might not.

Additional sections

There are profile sections within LinkedIn that you may not have taken the time to update or do not even know are available. LinkedIn does not have the same single-page constraint that a resume does; if you have information that may increase your chances of landing a job, include it.

Featured content

The Featured Content section is near the top of the page and allows you to add articles you have published, projects you have worked on, as well as your personal website(s). This should be a highlight of career accomplishments that can be shared publicly. A wonderful thing about the featured content section is it can easily be populated with items in the Accomplishments section, which we will cover next.

Accomplishments

Accomplishments are often forgotten about. The Accomplishments section can be used to add publications, patents, courses, projects, honors and awards, test scores, organizations, and causes. The Accomplishments section information can be added as featured content and displayed at the top of your page.

Experience

The Experience section is usually updated directly on LinkedIn by the individual who owns the profile; however, the details describing each experience are often lacking, especially if you have been in a role for an extended period. A good rule is to update your Experience section each time you finish working on a project or initiative. If you are looking for your first role as a DevOps engineer, try to highlight the experiences from your past roles that would apply to a role as a DevOps engineer.

Profile picture

Updating your profile photo is not necessary to get a job or be recognized; however, it will not hurt. The same person who takes your family photos or pictures of your pet could capture a headshot for you to use on your profile. Profiles that have a profile picture are ranked higher in the search algorithm, resulting in more recruiters and hiring managers seeing your profile.

In the next section, we will discuss the power of skill endorsements, as well as skill exams.

Skill endorsements

Skill endorsements can be added by you and validated by any of your first-degree connections. The more individuals who validate you for a job, the higher the likelihood you will be found when someone searches for the skill. If you are looking for a job as a DevOps engineer, getting endorsed for skills such as AWS, Python, and DevOps can help you get noticed. To further increase the likelihood of being found based on a specific skill, you can take skill assessments.

LinkedIn skill assessments are professionally written exams that you can take for specific skills; when passed, you receive a badge on your profile telling everyone you have passed the skills assessment. This endorsement again raises your rank in the search algorithm and increases your chances of being noticed by recruiters. The final thing we will discuss is interacting with other professionals within your industry.

Share, like, and comment

Interacting with other DevOps professionals has proven to be the most powerful means of connecting and building relationships on LinkedIn for me. In *Chapter 5, Building Your Network,* we will cover strategies to help build relationships on LinkedIn. Every time you share, like, or comment on content, the likelihood of you getting noticed or showing up in a potential employer's feed increases.

In this section, we went through strategies to help increase the likelihood of being noticed on LinkedIn; in the next portion of this chapter, we will go through strategies for ensuring your resume is ready to present to hiring managers.

Updating your resume to match the career you are after

Even if you are not actively looking for a job, it is good practice to have an updated copy of your resume ready to dispense, as the perfect opportunity can come about at any time. In this section, we cover tips to improve your resume, which will lead to more callbacks and, hopefully, interviews.

Regardless of how skilled you are or how much experience you have, your resume should fit on a single page. If Elon Musk can fit his experience on one page (`https://novoresume.com/career-blog/elon-musk-one-page-resume`), so can you. The purpose of a resume is to allow recruiters and hiring managers to gain a quick picture of a candidate in under a minute. Some companies implement a computerized layer in the process that scans your resume for keywords and verifies requirements prior to it being passed along to a human. Let's look at what should be included in the six sections of a resume, starting with contact information.

Contact information

Contact information should include at a minimum your name as it appears on your online profiles, email address, and phone number. Including your location and personal profile information is recommended, especially if you have just updated it. You are not encouraged to include a photo in your resume, as recruiters believe this can add bias to the selection process. The next piece of vital information is your objective.

Objective

This is the section that describes what you have to offer, and what type of position you are hoping to land. The objective is short and concise, no more than a sentence or two. Depending on your preferences, you can write a general objective, such as the following:

DevOps leader with 20 years of experience specializing in cloud-native security, and Kubernetes. Looking to join a fast-paced organization with a strong engineering culture.

One thing to notice about the preceding objective is the overuse of keywords to put emphasis on the skills you bring to the role you are searching for. The alternative to this is an objective targeting a specific position, such as the following example:

DevOps leader with 20 years of experience specializing in CICD, security, and Kubernetes, looking to join company xyz as a Lead DevOps Engineer.

The latter approach is a better option as it allows you to cherry-pick keywords to include, as well as to make it more personal by using the exact job title and company name. On the other hand, you may want to bulk apply to as many jobs as you can; in those cases, you should always have an updated resume with a general objective you can use. A keynote about the objective is it should include keywords that you want to be noticed, both by an automated checker and a human. In the preceding example, we used several keywords, including years of experience. Next, we will discuss the largest portion of your resume: your work experience.

Experience

Experience should be listed in reverse chronological order, with your most recent at the top of the section. Each position listed should include five key pieces of information: **position**, **employer**, **start date**, **end date**, and **accomplishments**. All information should be as accurate as possible. Always assume your information will be verified; even an honest mistake could cost you a job. Accomplishments should be specific and contain quantifiable data.

Worked with application teams to significantly increase availability for customer-facing applications

The preceding example is a friendly conversation starter; however, it will annoy a recruiter or hiring manager as it lacks clarity and leaves the reader wondering what was accomplished. A single poorly written accomplishment could get your resume put in the rejection pile if you are applying for a highly competitive job. A better way to write this would be as follows:

Implemented geographic redundancy and automatic failovers in AWS for app x, which resulted in application availability increasing from 98.7% to 99.9%.

The preceding example gives clarity and specific details about what was accomplished. This type of detail gets your resume moved to the interview pile instead of the rejection pile.

A common challenge for new DevOps engineers is that they lack experience. Not to worry, you just need to document the experiences that have prepared you for your first role as a DevOps engineer. If your current position is as a software engineer, make sure you focus on the accomplishments that are relevant for a role in DevOps by focusing on the relevant tools and principles. Something like the following would fit well:

> **Software Engineer | Company XYZ | May 2020 - Present**
>
> Integrated the end-to-end testing framework, Test Café, into Jenkins CI pipeline, resulting in a 30% reduction in time team spend testing locally.
>
> Worked on APIs utilizing AWS services, which currently serve over 20,000 requests per month.
>
> Contributed my Test Café method to the Jenkins Global Pipeline Library inner source project so other teams could leverage it.

My advice for readers is this: experiences included on your resume do not need to be paid roles. If you are a main contributor to an open source project, include that in your experience section, especially if you are early in your career. This will get you noticed and land you an interview. A common misconception is that unpaid experience and volunteer work should only be brought up during an interview.

Next, we will cover both the Skills and Certifications sections.

Skills and Certifications

Your skills and certifications should be a copy of what can be found on your LinkedIn profile; you have less space, so choose your most relevant skills and certifications to include. In *Figure 4.4*, bar diagrams are used to represent my competency. One thing known to raise questions is when your resume shows top skills that do not reflect the story your experience tells. Saying AWS is a top skill when none of your accomplishments for previous roles reflect that is a red flag for recruiters and hiring managers. If your experience with AWS has come from a large side project you have worked on, you need to include the project on your resume.

Education

I really started to feel old while researching this topic. A four-year degree used to be a requirement to get into a software engineering/IT-related role. As you progressed in your career, it became less and less important until it was a non-topic, unless you were looking to move into a leadership role that required a master's degree. Now, for a lot of individuals, it started out as a non-topic because they were self-taught, or went through a coding boot camp and gained real-world experience through internships. This is a step in the right direction. As a hiring manager, I am less concerned that you have a four-year degree and more concerned that you have a mindset of continual learning.

Any education you have should be listed on your resume; some roles have minimum education requirements that you will be automatically disqualified from if they are not found on your resume. In *Chapter 5*, *Building Your Network*, we will discuss how it is possible to bypass this requirement by networking with the correct people.

At this point, we have covered the required sections of a resume. The following is an example that visually demonstrates the various sections:

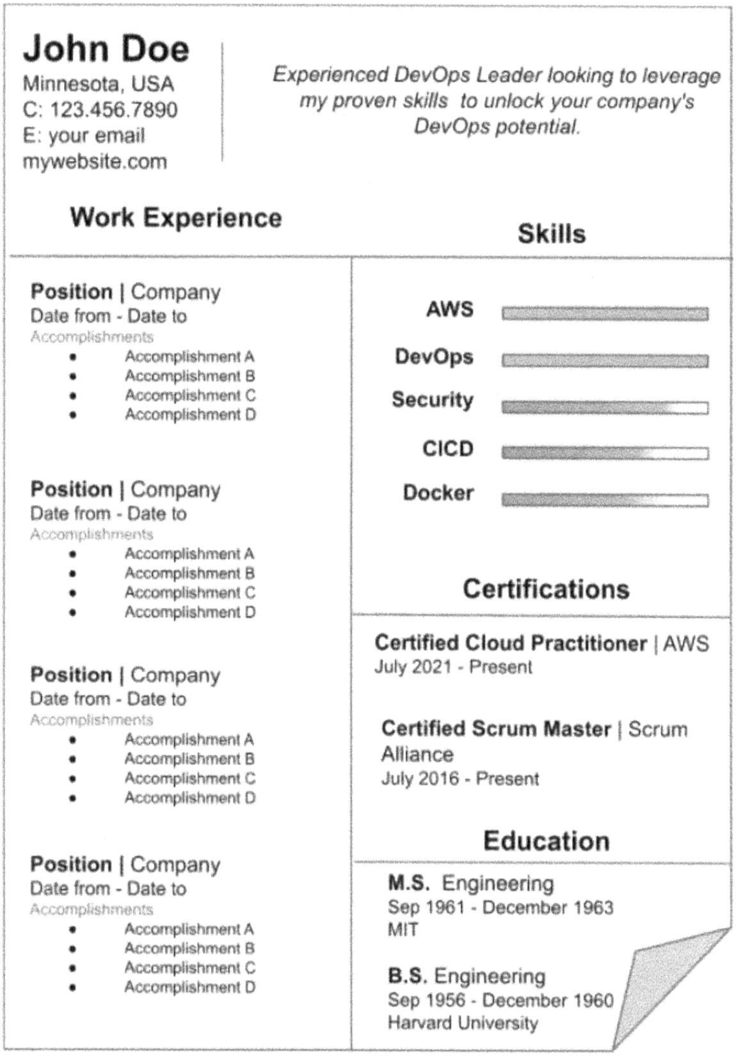

Figure 4.4 – A resume example

The following are two sites that can help you create amazing-looking single-page resumes:

- Novorésumé: `https://novoresume.com/`
- Resume.io: `https://resume.io/`

In the next section, we will discuss the importance of a personal web page and how to create one.

Updating and or creating your personal web page

Your personal web page should be an extension of your resume and LinkedIn profile, a place where you can expand on topics you are passionate about and projects you are interested in. The web page should allow you to discuss and divulge your personal interests and hobbies to humanize yourself to potential employers. If you already have a personal web page, you can skip the next how-to section on creating one using GitLab Pages.

GitLab Pages tutorial

Creating a static web page has become a simple task that costs no money, and not a whole lot of time. Both GitHub and GitLab offer free static site hosting. In this section, we will go through how to create a site using GitLab Pages.

Prerequisites: You need to be registered with a free account on GitLab.

1. The first thing we are going to do is install hexo, a node-based website framework:

    ```
    npm install -g hexo
    ```

2. Next, log in to GitLab (https://gitlab.com/) and navigate to https://gitlab.com/natejswenson/dcr-demo. Fork dcr-demo into your workspace.

    Clone the repository you forked onto your local machine, cd (change directly), into the project folder directly, and run the following commands:

    ```
    npm install
    hexo server
    ```

3. Open a browser and navigate to localhost:4000 to view the current site.

4. Make changes to personalize the site to make it your own and then push it back to GitLab:

    ```
    git stage .
    git commit -m "my commit"
    git push
    ```

Navigate back to GitLab to the project you just pushed, and click on **Settings | Pages** to view the URL where your site is published.

If you followed along, you have installed the necessary modules onto your local machine to do development using the Hexo framework, forked a repository into your user space, cloned the repository to your local machine, made changes, and pushed the changes back to GitLab where your site is published. Nice work. In the next portion of this section, you will be given the minimum information that should be included on your personal web page.

Sections to include on your personal web page

Development has always been easy; creating content on the other hand is a struggle. The following are what I have found to be effective sections on a personal web page.

Contact

The contact section should include your email, LinkedIn, GitLab, GitHub, and any other professional profile you want individuals visiting your site to be aware of.

Introduction

Introduce yourself to potential employers, both professionally and personally. This is your chance to describe who you are at work and outside of work, as well as to express your passion for the field of DevOps.

Blog

This section is optional; if you enjoy writing it should be included, as your writing will be of interest to potential employers. If you are not a writer, do not start blogging just to have it on your web page. I did this, and it was not fun for me or my readers; I was not passionate about what I was writing about, so it was not my best work. In this case, my time was better spent working on some Alexa Skills and adding those to my website.

Projects

This is another opportunity to show potential employers your interests and what you spend your time doing outside of work. A few examples include the following:

- GitHub projects
- GitLab projects
- Smart device skills
- Custom home automation
- School projects (especially Capstone)

- Speaking events
- Presentations you have given

Resume

Your resume should be visually displayed or downloadable from your site as a PDF. This reduces the back and forth between you and interested parties.

In summary, your personal website is an extension of your LinkedIn profile but allows you to add your own branding to it.

Leveraging Twitter and other social profiles

Whether you are a social influencer of technology, or a college student hoping to land an internship, it is beneficial to have a social presence that spans several platforms. In previous sections, we covered how to set up your LinkedIn page in such a way that you will get noticed by recruiters, as well as how to get started with your own web page. In this section, we will take this a step further and discuss how you can supplement your LinkedIn and personal web page by posting on Twitter or writing articles for Medium.

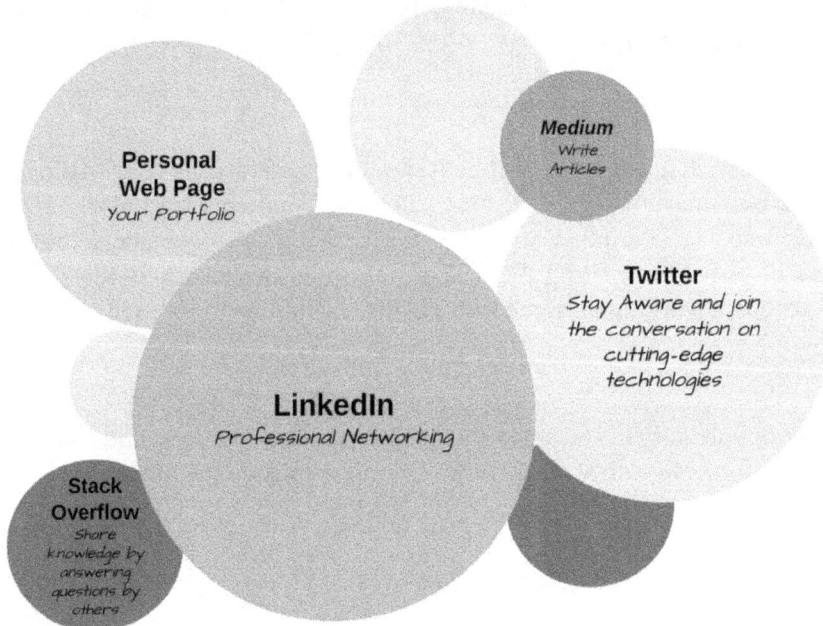

Figure 4.5 – Social profiles

We will start by discussing Twitter and how you can leverage it to get a job as a DevOps engineer.

Twitter

Twitter use for technology professionals falls into two categories: consuming information and sharing information.

Twitter for consuming information

Twitter is a wonderful place if you are looking to stay current on the bleeding edge of technology announcements. On Twitter, you can follow individuals such as Gene Kim or Martin Fowler, as seen in the following figure:

Gene Kim
@RealGeneKim

WSJ bestselling author: Unicorn Project! DevOps researcher/enthusiast. Coauthor: Phoenix Project, Accelerate. Host of The Idealcast. Tripwire founder. Clojure.

UT: 45.527981,-122.670577   itrevolution.com/idealcast/
Joined January 2009

1,618 Following   **52.2K** Followers

Martin Fowler
@martinfowler

Author, speaker, and general loud mouth on Software Development. Works for Thoughtworks. Also hikes, watches theater, and plays modern board games

Boston   martinfowler.com   Joined October 2008

261 Following   **329K** Followers

Figure 4.6 – Gene Kim and Martin Fowler on Twitter

On Twitter, you can also follow news outlets and companies, such as Stack Overflow or ZDNet, as seen in the following figure:

Stack Overflow
@StackOverflow

Stack Overflow empowers the world to develop technology through collective knowledge.

New York, NY   stackoverflow.com   Joined April 2010

99 Following   **135.3K** Followers

ZDNet
@ZDNet

Where technology means business

USA | UK | Asia | Australia   zdnet.com   Joined April 2007

204 Following   **464.4K** Followers

Figure 4.7 – Stack Overflow and ZDNet on Twitter

Another use of Twitter is for sharing information with your followers.

Twitter for sharing information

If you become a fan of Twitter and enjoy using it, you may want to try and increase your followers. An effective way to do this is by sharing original content or sharing content posted by someone else.

If you become a popular and highly followed account on Twitter, you'll be more likely to get noticed by recruiters and individuals who are hiring.

Twitter does have a limitation of posts, or tweets, having a length of no longer than 280 characters. If you need a forum that allows for more information, you may want to consider Medium.

Medium

Medium, according to `https://medium.com/`, is an open platform where readers find dynamic thinking, and where expert and undiscovered voices can share their writing on any topic. Medium is a wonderful place to start your career as a writer, as anyone can write for it.

The key to finding success on Medium is building views of your article. This can be done by cross-posting your Medium article on other social sites, such as LinkedIn and Twitter.

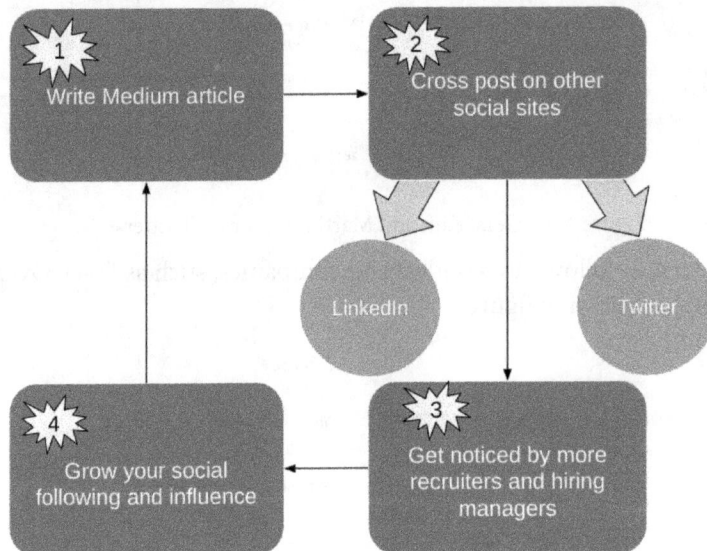

Figure 4.8 – Using Medium and other social sites

In the preceding diagram, you have just written and posted an article on Medium but have struggled to get people to click on and read your article. You decide to post your Medium article on your LinkedIn and Twitter profiles. This results in getting additional clicks for your Medium article. You also begin to receive messages from recruiters and others congratulating you on your article.

There are many other social media sites that can further help you grow your online presence, which will increase your chances of connecting with individuals who can help you land a job.

Summary

In this chapter, we covered all the necessary changes you need to make to ensure you are presenting your best self to potential employers. We discussed the importance of having a complete and professionally written LinkedIn profile, as well as giving suggestions on how to improve the likelihood of being noticed by potential employers. Next, we covered how to update your resume in such a way that essential information is seen quickly, both by automated systems and humans. We then covered the importance of having a personal web page, went through a tutorial on how to create a Hexo web page on GitLab, and the sections that should be included in your personal web page. Finally, we covered other social sites, which are not required but can increase the likelihood of getting noticed by recruiters.

In the next chapter, we will discuss the importance of networking, and how to do it on LinkedIn and at conferences.

5
Building
Your Network

My current job is great. I enjoy the work I am doing, I am learning every day, and I have a great manager; a true triple-threat job. You may be thinking I had to work extremely hard to get this job. In a way I did, but not in the way you are probably thinking. I did not work hard to prepare for the interviews, as most of the interviews happened without me even knowing I was being interviewed. This was because I had built a relationship on LinkedIn with the hiring manager. For several months, we had been discussing different job opportunities within the organization he worked for, and at the same time, he was getting to know more about my skills and career goals.

In this chapter, we will discuss how to build relationships to help your career move in the right direction. The saying *It isn't what you know, it's who you know* is partially true. If I were to reword the saying so it was completely true, I would say *Getting noticed for your skills by the right people is the key to success*. In this chapter, the following topics will be covered:

- LinkedIn the right way
- Building lasting connections, online and offline
- Quality over quantity
- Networking – conversation starters

We will start by discussing how to use **LinkedIn** to get noticed by recruiters and people hiring DevOps engineers, followed by how to build lasting relationships with the people who end up noticing you, both on LinkedIn and at meetups and other networking events. Afterward, I will explain why having a few connections with whom you are able to build relationships is better than having many connections. Finally, I will close this section by giving a few conversation starters for introverts, like myself, who are reading this book.

LinkedIn the right way

LinkedIn started as a professional network; it has since expanded to be one of the leading job search sites. A few key statistics about LinkedIn to get you excited are shown in the following figure:

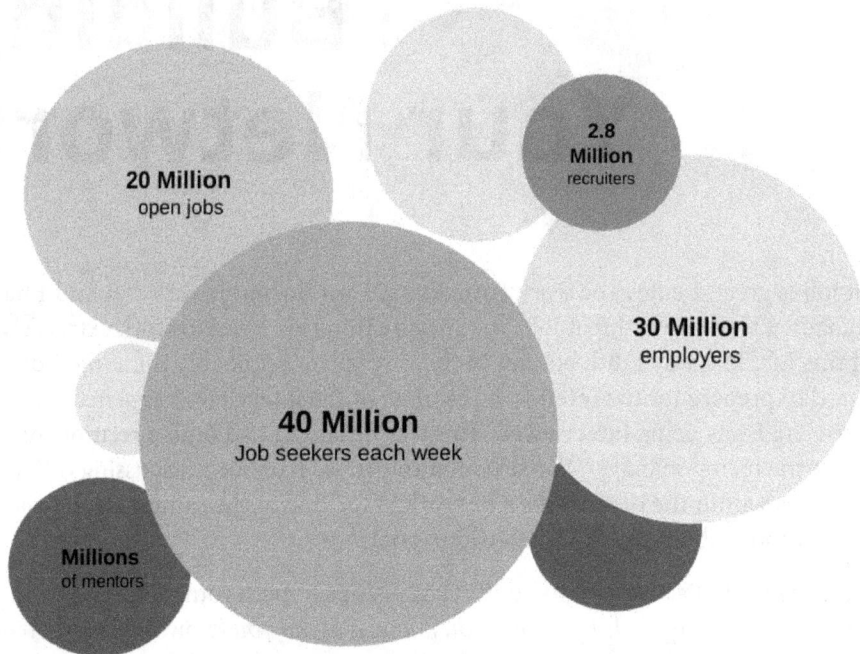

Figure 5.1 – LinkedIn infographic (2021)

Those are some large numbers. This section will guide you through how to correctly leverage them. The first step toward building a respectable reputation on LinkedIn is getting noticed.

Getting noticed

Getting noticed on LinkedIn can seem impossible, with the number of users approaching 1 billion. One of the easiest ways to get noticed is to change your security settings to be an open networker. After you have opened your profile up, it is time to start following and interacting with other DevOps professionals. If you lack DevOps connections, an effortless way to ensure you start seeing posts about #DevOps, #AWS, #DevOpsJobs, #CI, or anything else, is to follow specific hashtags. This will show not only your first-degree connections' posts but also second- and third-degree connections' posts. We will use the following diagram for the remainder of the section:

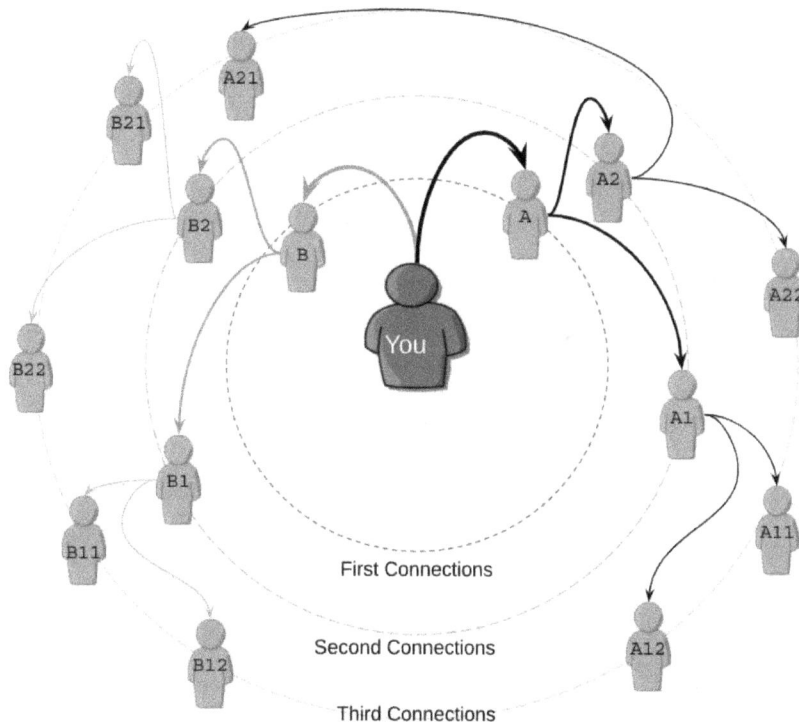

Figure 5.2 – LinkedIn connections

Pretend you have two friends, A and B; these are your first-degree connections or the ones you are directly connected with. In our example, each of your first-degree connections has two connections, and each of your second-degree connections has two connections.

This translates to 14 connections in your extended network. These numbers increase exponentially as connections rise; with 50 first-degree connections, and each of your first-degree connections having 50 connections and each of your second-degree connections having 50 connections, the total connections in the first through third degrees amounts to an astounding 1,277,550.

One way to get recognized by several connections is by posting about something you are learning about. Connections begin to comment on it and like it, and it ends up being seen by a recruiter who is looking for DevOps engineers with Kubernetes experience, as shown in the following diagram:

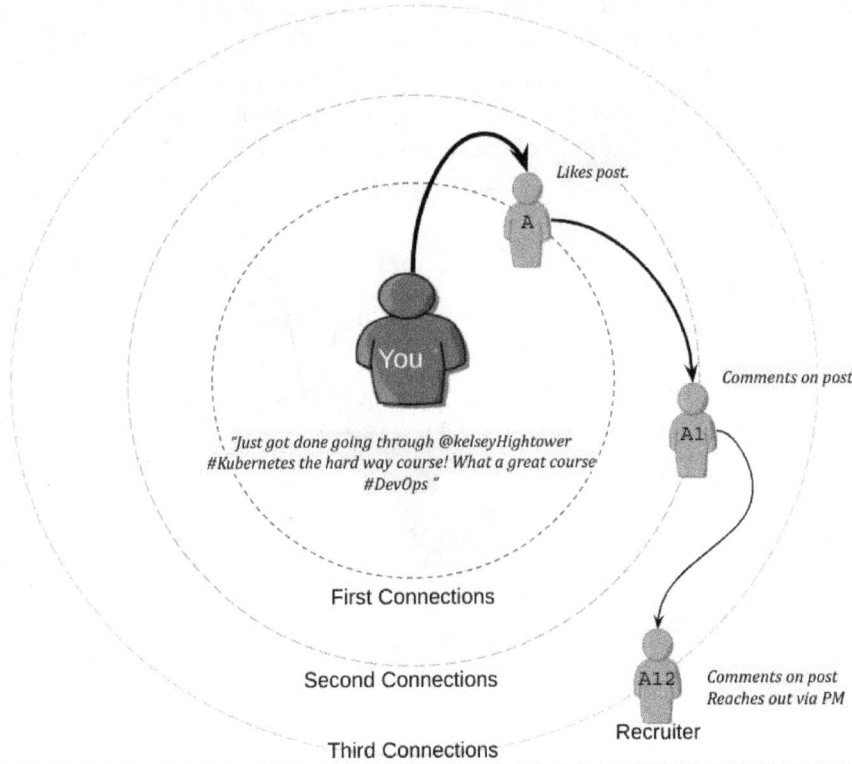

Likes post.

A

You

"Just got done going through @kelseyHightower #Kubernetes the hard way course! What a great course #DevOps "

Comments on post

A1

First Connections

Second Connections

A12

Comments on post Reaches out via PM

Recruiter

Third Connections

Figure 5.3 – A LinkedIn post leading to a third-degree private message

This may sound unlikely, but it happens to me time after time. After the recruiter has reached out to you via a private message, you can add them as a connection. This is a suitable time to bring up the previous lesson from *Chapter 4, Rebranding Yourself,* on ensuring your LinkedIn profile is presentable; you do not want recruiters viewing your incomplete or messy profile.

Another fantastic way to get recognized on LinkedIn is by commenting on and/or liking someone else's post. In the following example, *B1*, a second-degree connection, posts about starting a meetup in the area where you live. One of your connections has already liked the. It shows up in your feed, so you like it, and comment on it as well. At this point, follow up with a personalized invite to connect, expressing your interest in the meetup:

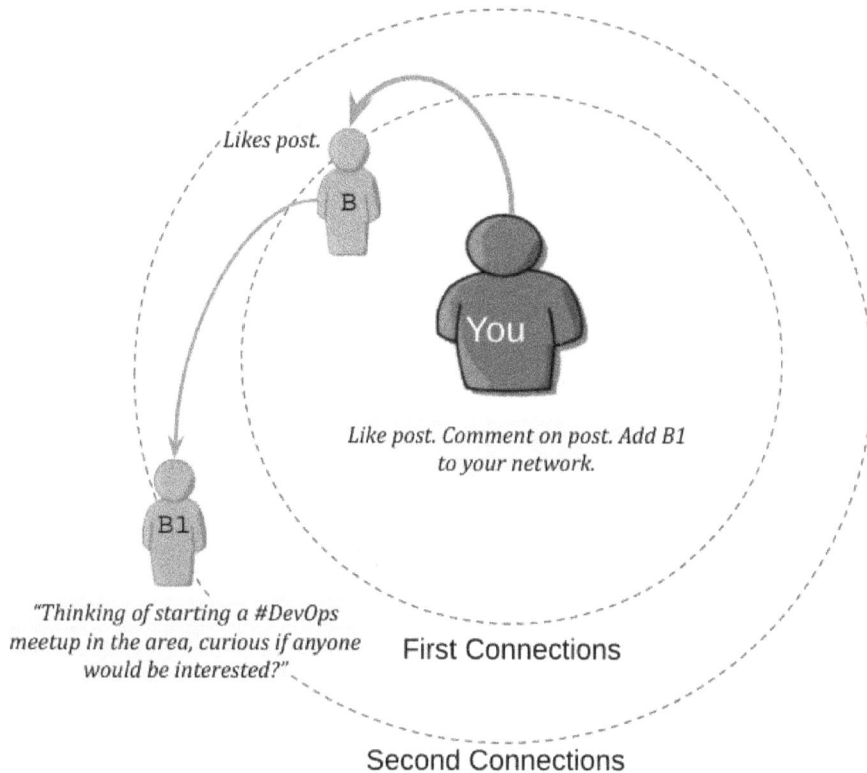

Likes post.

B

You

Like post. Comment on post. Add B1 to your network.

B1

"Thinking of starting a #DevOps meetup in the area, curious if anyone would be interested?"

First Connections

Second Connections

Figure 5.4 – Adding a second-degree connection to your network

To recap this section, the best way to get noticed on LinkedIn is by interacting with connections, even if they are not directly connected to you. In the next section, we will discuss how to build lasting connections.

Building lasting connections, online and offline

You may get fortunate and be offered a job after one interaction with someone. This is not something that has happened to me; it has always been a long game when it comes to payoffs from relationships. There was an instance where a recruiter reached out to me about a job. At the time, I was not looking for a job change. We stayed connected on LinkedIn and met for lunch a few times a year. Three years later, a friend and former colleague lost his job after his department was restructured. I sent his resume over to the recruiter; within 2 weeks he was starting a well-paying contract gig that was able to hold him over while he searched for a more permanent position.

In this section, we will discuss ways to build lasting connections, both in a virtual setting and in person. We will begin by discussing how to build connections online, or in a virtual setting.

Building connections in virtual settings

In DevOps circles, remote work and virtual relationship building are things that were happening long before Covid-19 turned our world upside down. Companies such as **GitLab**, **Atlassian**, and **PagerDuty** are remote-friendly, and team members have amazing rapports with one another. I have been working remotely for 5 years, and believe the same practices used to ensure a strong bond among team members can be applied to any type of professional relationship.

The following is the visual setup for the scenario of building relationships by being personal, helpful, and consistent:

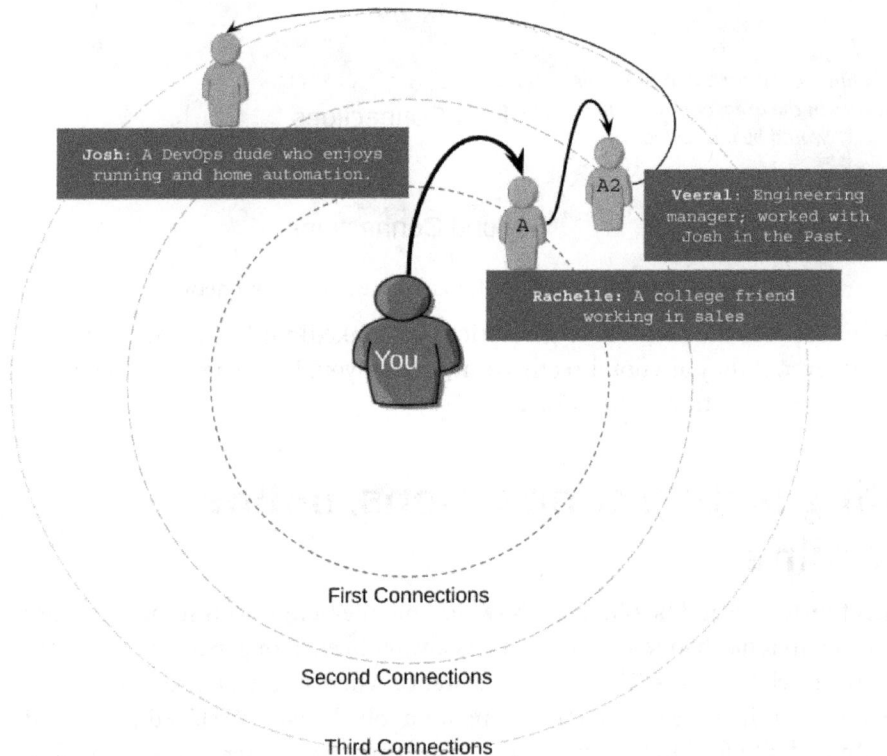

Figure 5.5 – Setting up a scenario in LinkedIn

Building connections in a personal setting

There is a line that you must not cross, obviously, but people love hearing about your pet and your activities outside of work. Take the preceding example. Let's assume Josh posted about completing a marathon. Rachelle likes it, and it shows up on your news feed. You enjoy running as well, so you decide to PM Josh: "*Congratulations on completing your marathon, what an accomplishment! I am a huge fan of running as well.*" Messaging Josh has allowed you to connect on a personal, not work, level, which can help build stronger relationships.

Help your connections

Helping your connections is the most important tip on my list. After you sent Josh the PM, he looked at your profile and noticed you were both in DevOps, so he added you to his network. He ends up posting about having trouble with a side home automation project he is working on involving **Raspberry Pi**. You happen to have a similar setup, so you decide to share your GitHub link with the code you used in your setup. A few weeks later, you end up posting about an issue you are having with AWS, asking your network for help. Josh sees your post and comments on it, as well as PMing several solutions he has used in the past. His ideas end up being great; you ask if he would like to hop on a video call and explain a few more things in detail.

Consistently engage with your connections

At this point, you and Josh have a decent relationship. To keep it like this, you are going to need to stay engaged with him. Set some cadence where you check in with each other and see how things are going.

In closing this section, I will leave you with several virtual meetups that can help you discover new professionals with the same interests as you; first, **All Day DevOps** (**ADO**) `https://www.alldaydevops.com/`. If you search for online DevOps conferences or virtual DevOps meetups, you will be amazed at how many there are. I encourage you to attend a few and figure out which ones will be of the most benefit to you.

In real life

Covid-19 brought a new term into the workplace, **in real life** (**IRL**), and it is a clever way to describe meetups in person with colleagues and coworkers, or potential colleagues and coworkers. Whether you are an introvert or extrovert, navigating IRL professional relationships is complicated, which is what we will address in this section. The biggest benefit from IRL events is the ability to get one-to-one time with people you want to connect with better.

Next, we will discuss some attributes that are important in virtual relationships but are much more noticeable when dealing with IRL situations.

Authentically show others your true self

The best advice I was given by my mentor, which I would like to pass on to my readers, is to *be authentic*. My mentor was referring to the way I engage with people. Some people prefer a more intimate one-to-one conversation, while others feel more comfortable in a large group conversation. Put yourself in a situation where you can present yourself in an authentic manner.

If you try to force yourself into a situation where you must act in a way that is not natural or comfortable for you, the conversation will not go smoothly. The individual you are trying to build a relationship with will realize you are not being yourself and will be wary of continuing to associate with you.

Engage actively with your network

If you have been focusing on learning a specific topic for an extended period, volunteer to be present at an upcoming event. This will have multiple benefits. First, it will ensure everyone who attends the meetup knows who you are, which is great if you are in the market for a career change. It also allows you to add *a presenter* to your LinkedIn profile, which is another terrific way to get noticed. Finally, if you are early in your career, it is a wonderful opportunity to get practice presenting, a skill that will be useful for the rest of your career.

Another way to be engaged is to participate in back-and-forth discussions with the presenter directly after a presentation. Do not be argumentative. That will not build relationships; however, ask questions and sound interested. This will get you noticed.

Hold yourself to high professional standards

This one is common sense, but it is amazing how many people forget this simple rule. After events, there is always a happy hour. It is easy to forget you are at an event to help build your career, and instead think you are out having fun with your friends. There are two simple rules to remember: keep the conversation professional, and do not overindulge in beverages. It is hard to come back from a poor first impression.

In summary, to be successful at in-person events, bring your authentic self to the event, be engaged at the event, whether you are in the audience or are the presenter, and finally, be professional.

It is important to end this section with a list of the six personal attributes that will make you more successful at building relationships, which are visually displayed in the following graphic:

Figure 5.6 – Attributes required for building relationships

Reiterating what is shown in the preceding graphic, you must try to be helpful, consistent, engaged, authentic, professional, and personal.

In the next section, we will cover the importance of quality of connections over quantity of connections.

Quality over quantity

Two people make a post on LinkedIn at the same time regarding the same topic. Connection A has 10,000 connections, while connection B has 800 connections. After a day, connection A's post has two likes and no comments, while B's post has 14 likes and 22 comments.

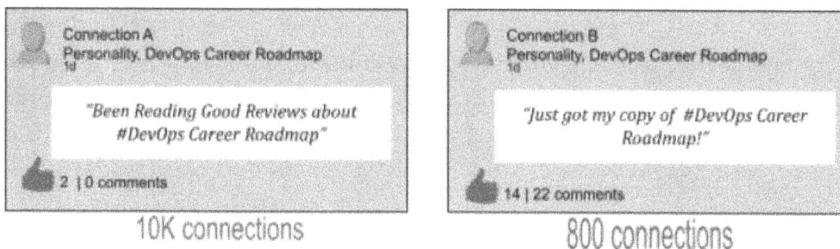

Figure 5.7 – Looking toward quality

When I first came across a situation like this, I was confused; was connection A doing something wrong, or was connection B? Let me break it down.

Connection A is focusing on growing their network, paying little attention to who is being added to their network. The number of connections is a vanity metric; it looks good at first glance but if you look deeper, it is meaningless.

A more useful measure is how many times, and at what frequency, your connections interact with you. By this measure, connection B is definitely doing something right, but what?

Well, if you have been reading this chapter, connection B has been doing all those things we have talked about. They have made a conscious effort to stay engaged with their connections, help others out, and get personal. Most important, connection B is always their authentic self, online and IRL. The following are questions you can ask to determine whether you should add a connection to your profile. If you answer yes to one or more, go ahead and add them; if not, do not add them. This is something that ensures each of your connections is someone you know, someone who interests you, or someone who can assist you:

- Do I know this person?
- Do I feel drawn to this person's business, career path, or professional outlook?
- Do I have a direct need this person can fill?

The takeaway from this section is a culmination of behaviors that have been described throughout this chapter: it is better to have a few connections with who you have robust relationships rather than many connections who are nothing more than numbers. More simply, put *quality over quantity*. In the last section of this chapter, we will discuss conversation starters you can use when at networking events.

Networking and conversation starters

There will be a few reading this book who, like me, really got the short straw when it came to being good at small talk. There will also be those who would prefer to skip the small talk altogether and get to some deeper, more relevant discussions. In this section, you will learn a few useful techniques for facilitating a networking event, as well as conversation starters and redirects that can be used when you find yourself at a loss for words or feel it is time for another topic. Let's start with one that can be used to facilitate agenda-less meetings, **Lean Coffee**.

Lean Coffee

Lean Coffee is a way to structure a meeting/event in a way that allows participants to vote on topics they would like to discuss. I have used this in book club discussions I have hosted as well to make a meeting with an over-stacked agenda much more productive.

The following figure is a visual depiction of the essential items needed to facilitate a Lean Coffee meeting. It is also likely the artwork I am most proud of, as I am not an artist:

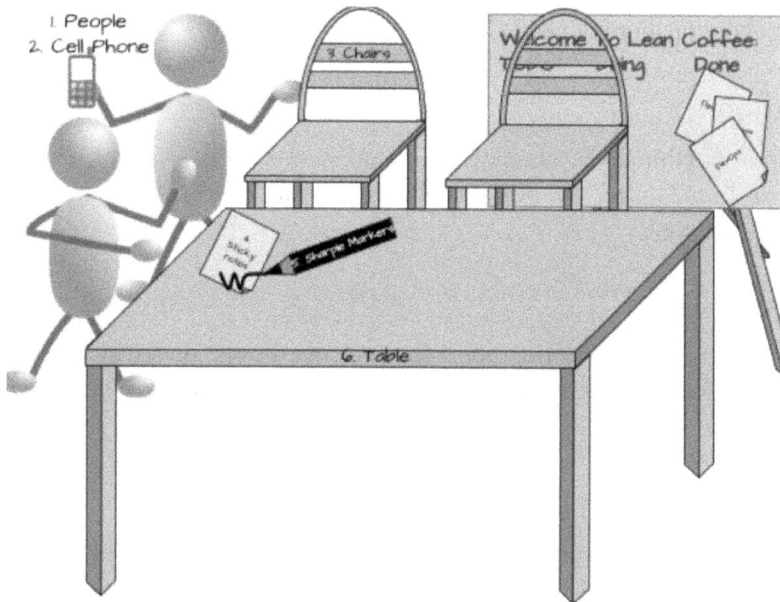

Figure 5.8 – Lean Coffee essentials

The minimum requirements to run a Lean Coffee event include sticky notes, markers, a smartphone or other timing device, a table, chairs, and most importantly, people. Next, we will talk about facilitating Lean Coffee. We will start with a visual depiction followed by detailed descriptions of each stage:

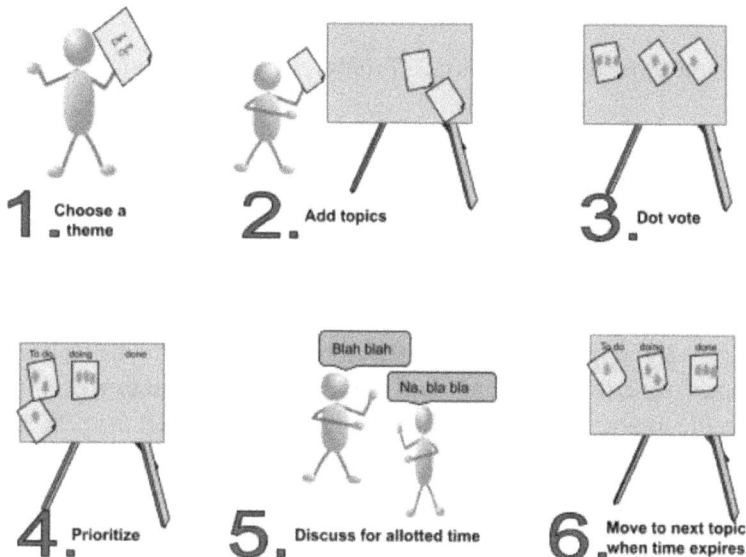

Figure 5.9 – Lean coffee "how to" visual

Now, let's talk about facilitating Lean Coffee.

Choose a theme

The theme can be anything, such as DevOps, security, or Agile in DevOps. They are all good themes for Lean Coffee. The purpose of them is to drive participants to add topics around a common theme.

Add topics

Participants add topics related to the theme they wish to discuss on sticky notes in as few words as possible. Topics, if the theme was DevOps, could include CICD tools, DevOps culture, and Docker.

Dot vote

Participants are given a set number of votes to add to topics they would like to discuss. I have found simple Sharpie dots on the sticky notes work well. If you would like to be more formal, you can do it on a whiteboard.

Drum roll, please.

Topics are sorted in order of most votes to least votes. The topic with the most votes is discussed first.

Discuss

Start the timer for a given time – usually, 3–7 minutes is sufficient, and begin discussing the first topic.

Thumbs up/thumbs down

Sticking with the democratic nature of Lean Coffee, after the time runs out for a topic, additional time can be given if most people wish to continue the discussion; otherwise, move on to the next topic.

You can be creative and apply the same concept to a virtual meeting using breakout rooms and the whiteboard functionality. At the end of the session, it is up to you and your group whether you decide to post notes on the Lean Coffee discussion. In the past, I have used GitHub pages to track outcomes of Lean Coffee, as well as upcoming dates.

Next, I would like to discuss lightning talks, which work great as both virtual and IRL events.

Lightning talks

Lightning talks are short presentations, usually 3–5 minutes in length, given in a forum where multiple presentations are given back to back. I absolutely love listening to lightning talks, especially in a venue where there are other individuals. Lightning talks introduce the audience to highly technical topics that leave the audience wanting more, which leads to some great post-presentation conversations between the audience and speakers.

Every lightning talk session I have attended led to conversations about things I did not even know I was interested in, with people I would have otherwise never met. It is possible to start an internal, recurring lightning talk series in your DevOps or engineering group. When I worked at *Optum*, we incorporated a lightning talk at the beginning of our DevOps community office hours calls. It got attendees engaged from the start of the call, and also led to great discussions.

Conversation starters

You do not have quality connections because you aren't having conversations with people at networking events, and you aren't having conversations with people at networking events because you do not know what to talk about. We have all been there, or at least I have been there more times than I care to admit. In the previous sub-sections, we discussed lightning talks and Lean Coffee events that were purposefully designed to inspire conversations. Unfortunately, most networking events consist of refreshments and people gathered in a space. The following are a few things that have helped me as I matured in my ability to network effectively:

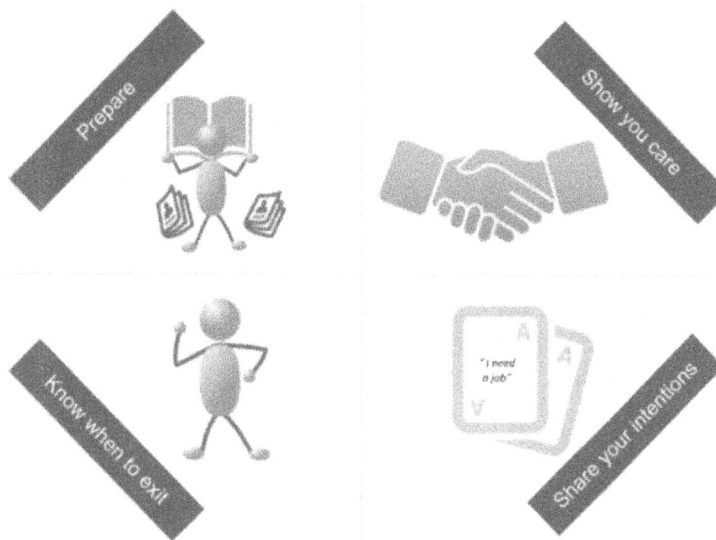

Figure 5.10 – Ways to improve your conversations

We will first discuss the importance of being prepared.

Come prepared

If you are going to a Docker meetup, I would binge-watch a few videos on Docker, as well as reading some recent articles, to ensure you are able to have effective conversations with individuals who are at the event.

Get personal

Ask questions that will give you more information to continue the conversation. An example of a poor initial question would be, *Hi, how are you doing?* This gives you no information to set up your next question and will lead to a short-lived conversation. A better conversation would be, *Hi, I see from your name tag you work at Amazon. What is it that you do there that has you interested in containers?* This is a good question. First, it cannot be answered with a single-word response; second, the response you get will help set up your next question; and finally, this question makes it seem as though you are terribly interested in getting to know them. Let's pretend the individual is the manager of a team focused on AWS **ECS (Elastic Container Storage)**and is interested in Docker to ensure they stay at the bleeding edge of what is happening in the container space. The individual then turns the question on you, which brings me to the next point.

Be upfront with your intentions

If you are only at the event to network or connect with individuals who may help you land a job, state that. Being candid and forthcoming are traits many people admire, and few possess. The main point here is to be honest.

Know when and how to leave a conversation

If a conversation is going well, for instance, if after you stated your intentions, the individual seemed interested in learning more about you, you should spend as much time with the individuals as they need. Do not bow out of the conversation to move on to the next. You have laid the groundwork; now, it is time to build upon the foundation. The opposite of this is when the individual is not interested in networking for the sake of hiring someone now; thank the individual for their time, and leave by telling them to stay in touch via LinkedIn. Afterward, make sure to follow up and add them on LinkedIn with a message thanking them for the conversation; you never know when they may be in the market for a candidate.

Summary

In this chapter, we discussed several ways of networking, both virtually and IRL. We first discussed the best strategies for building relationships on LinkedIn. This included ways to increase the probability of you being noticed by others. In this section, you also learned ways to utilize your second- and third-degree connections to build relationships. The six traits that will improve the likelihood of successful conversations, virtually and IRL, were discussed. These consisted of being helpful, authentic, personal, consistent, engaged, and professional. We discovered that having connections that you interact with is more valuable than having many connections that you have no personal relationship with. Events that can help facilitate better conversation, such as lightning talks and Lean Coffee, were discussed next. Finally, we covered four methods that will drastically improve your ability to have meaningful conversations and build relationships at networking events. These consisted of coming prepared, showing you care, being upfront with your intentions, and knowing when to exit a conversation.

In the next chapter, we will cover mentorship: why it is important, and how to find a mentor.

6
Mentorship

In the field of **DevOps**, having a mentor will allow you to arrive at your career destination much quicker; it allows access to knowledge from someone who has been where you are. In this chapter, you will be given the motivation to find a mentor, along with the resources to find a mentor in a way that works with your personal preferences.

In this chapter, the following topics will be covered:

- The importance of mentorship
- The mentor-mentee relationship dynamic
- Choosing the correct mentor
- Mentors as references

The importance of mentorship

In this section, we will cover the benefits that a **mentorship** relationship can bring. This section will be broken down into four parts: the assistance and guidance toward setting achievable goals, career coaching, motivation, and career advice, as shown in the following diagram:

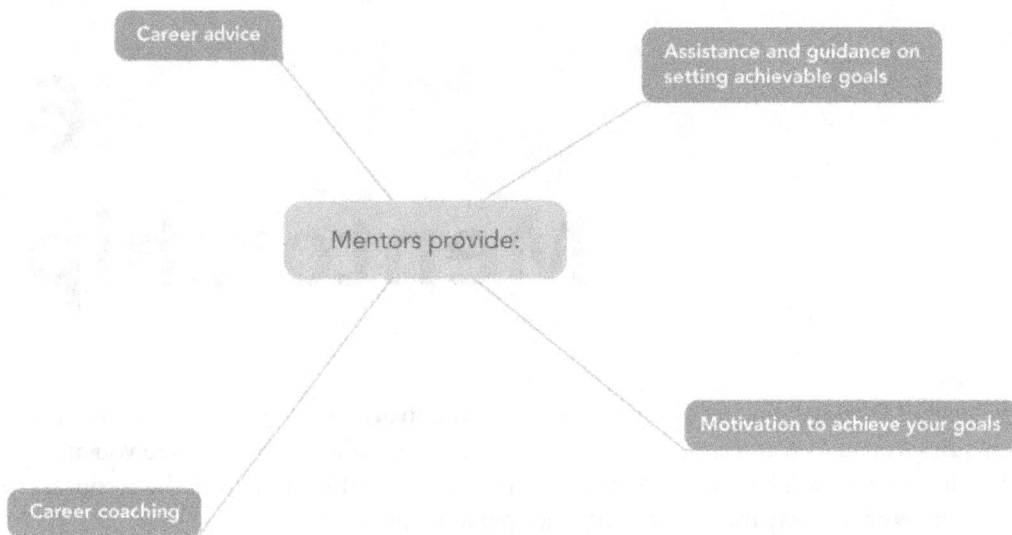

Career advice

Assistance and guidance on setting achievable goals

Mentors provide:

Motivation to achieve your goals

Career coaching

Figure 6.1 – The benefits of mentorship

We will start by covering how a mentor can help you set, plan, and reach your goals.

Assistance and guidance toward setting achievable goals

A mentor is going to provide guidance on how to achieve your goals. A mentor will help you ensure your goals are achievable, using milestones. The following is a true story of how my mentor helped to lead me down a path unrelated to my goal, which ended up being exactly what I needed to succeed.

> **Real Experiences: From DevOps Engineer to Technical Agile Coach**
>
> Leading a DevOps team had been my goal since starting a career in DevOps at *United Health Group*. My mentor helped surface the shortcomings that needed to be addressed if my goal was to become a reality. Presenting and demonstrating ideas to others had always been a struggle; also, being very timid and not great at voicing my opinion were all deficiencies I was unaware of until I had a candid goal-setting discussion with my mentor.
>
> The solution to overcome my shortcomings was transitioning from a behind-the-scenes technical individual to a coaching role. A technical coach's role was to help agile teams learn and adapt CICD (Continuous Integration and Continuous Delivery)best practices; in other words, a lot of demonstrating and voicing my opinion was going to be required.
>
> The first portion of my transition included being immersed in the world of agile, along with having 1:1 coaching sessions with my mentor. One of my first teams I was assisting was not using a CI server, had not migrated their code into GitHub, and had no automated tests or regression tests in place. These were things I was comfortable doing in my sleep; however, my role was not doing the work but instead coaching and guiding the team in the correct direction. It was terrible; however, slowly over the course of the first sprint, I started becoming more comfortable with myself and the team. It became quite fun, and also, my mentor and supervisor told me I was quite good at it.
>
> My journey as a technical coach lasted for two years before transitioning into leading the DevOps **Center of Excellence** (**COE**). It wasn't until much later that my goal to become a people manager of a DevOps team become a reality. Without the direction of my mentor, I likely would not have developed the skills that I needed to lead a team, nor would I have discovered the passion I had for coaching and training, which was a direction I, ultimately, decided to pursue further.

Often, a mentor is going to have an insight that you are unlikely to realize on your own. Before making any career-altering goals or changes, I encourage discussions with your mentor to get their input. If you have yet to find a mentor, continue reading this chapter for insights into finding one. Next, we will discuss the importance of coaching and training, which is another set of helpful skills a mentor offers.

The motivation to help you achieve your goals

After setting a goal, it quickly can become less enticing to work toward the goal due to the effort that is required. In situations such as this, a mentor will step in and reiterate why the destination is worth the effort. This is only possible if you and your mentor have a relationship that allows for open communications, that is, a relationship in which you feel comfortable discussing your doubts.

Your mentor is your cheerleader who wants to see you achieve your goal. To ensure you stay on the correct path, set up regular coaching sessions to discuss your progress and work on unlocking and enhancing your skills.

Career coaching

Training is the transfer of knowledge from a mentor to a mentee, while coaching is used by mentors to enhance a specific skill of a mentee. In the coaching sessions that you have with your mentor, it is your responsibility to drive the conversation and discussion in the direction you would like to go. This means being open with your mentor; if you are struggling with a specific concept or method, make this known to your mentor.

There are things that you need to bear in mind when being coached so that you do not get angry or become discouraged:

- A coaching session is not the same as training; a mentor will not explain a specific concept or method to you. Instead, they will push you in a direction they feel will help you better understand a particular concept.

- A coaching session is about you the mentee. Asking the mentor what they would do in a particular situation will be turned back on you. Remember that this is about how you are going to become better, not how or what your mentor knows or thinks.

- If you walk away from a coaching session with more questions than answers, it was a success! Take the questions you have and turn them into answers for yourself.

In the next session, we will discuss how mentors are there when you need a trusted individual's opinion.

Useful advice

A mentor is there when you need advice at any time. Whether it is about a project you are working on, or a general question about career progression, a mentor is there to help you traverse your career. Advice from a mentor is training in a less formal setting. When working on a project and an issue arises, your mentor might be the first person you message for their input.

Pro Tip: Advice from a Mentor Is Continuous Training

Each discussion you have with a mentor imparts a small piece of your mentor's knowledge onto you. When you ask for specific advice, it is like opening a mini training course from one of your most trusted publishers. Make sure you take full advantage of this opportunity if you get it.

In this section, we discussed the importance of having a mentor. Unlocking the potential benefits described means you must build a strong relationship that is built on trust and respect with your mentor. The sooner you feel comfortable asking your mentor for advice, the sooner you can unlock the potential of a mentor-mentee relationship.

In the next section, we will discuss the relationship dynamics between a mentor and mentee along with the various stages of the relationship.

The mentor-mentee relationship dynamics

You get to choose your mentor; however, your mentor must also choose you as their mentee, which is what makes the relationship unique and powerful. There is a core set of skills that is shared between both a mentor and mentee, as shown in the following diagram:

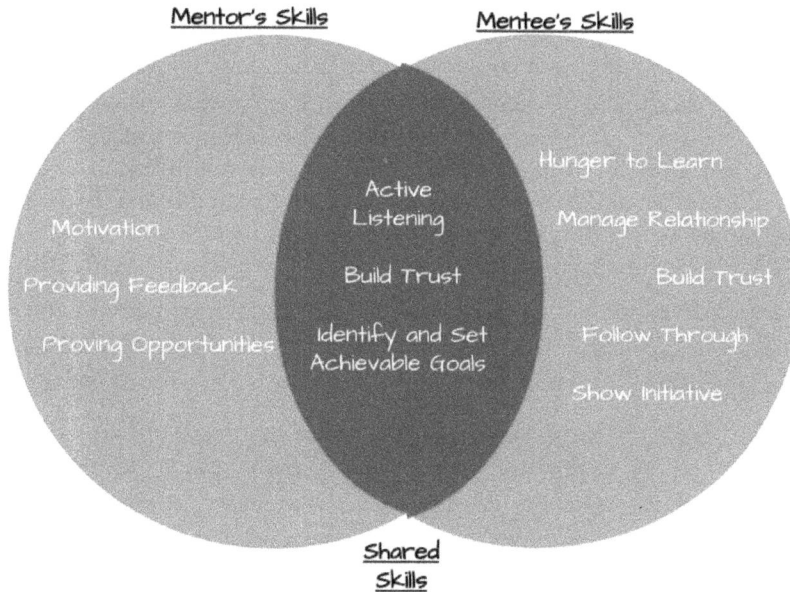

Figure 6.2 – The mentoring skill model

Active listening, **building trust**, and **goal setting**; all three of these can be tied back to respect. A mentor must respect his mentee, and likewise, a mentee must respect his mentor. There are three stages to a mentor-mentee relationship.

In the *first stage*, you begin seeking advice from someone you look up to. At this point, the specifics of the relationship have not been defined, and you officially do not have a mentor. At this stage, you are really testing the waters of someone you feel has the potential to be a good mentor. A good analogy for this is dating; you go on a date with someone you feel might be a good match for you before you ask the individual if they would like to become more serious.

The *second stage* is when you **define the relationship (DTR)**. In the second stage, you have officially asked someone to be your mentor, and they have agreed to your request. At this stage, you will likely have scheduled 1:1s and check-ins with one another.

It is important that you do not become complacent during this stage of your relationship, as the relationship is fragile and its infancy, you want your mentor-mentee relationship to continue to grow and mature. The following is another example from my personal career where I made the mistake of allowing my mentor-mentee relationship to get stagnant in stage two.

Real Experiences: What a Mentor Relationship is Not

Early in my career, I had several mentors I had officially asked to be my mentor. I got career advice from those individuals, and they provided me with guidance when I asked. The relationships were forced and awkward, which lead to very few 1:1 sessions and no coaching sessions. I was under the impression that this is just how mentor-mentee relationships were supposed to be.

The problem experienced in the preceding example stemmed from a lack of knowledge and understanding of how mentor-mentee relationships work. The destination for a mentor-mentee relationship is one that is more like a friendship; this is the third stage.

The *third stage* of a mentor-mentee relationship is where a strong bond grows between the mentor and the mentee. It is at this stage that trust enables the occurrence of effective coaching and 1:1 mentoring sessions. Depending on how deep your relationship with your mentor becomes, your mentor might begin considering you to be their protégé.

The following diagram is a graphical representation of the three stages of mentorship:

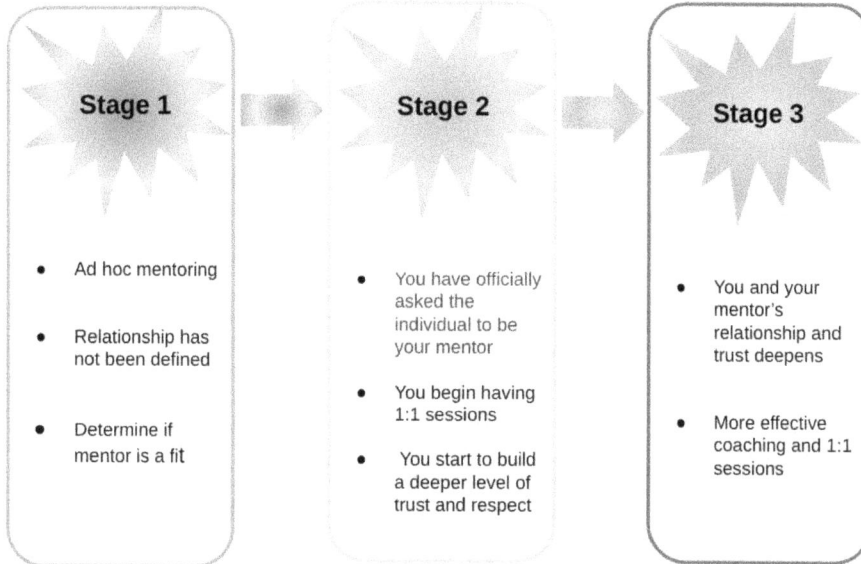

Stage 1
- Ad hoc mentoring
- Relationship has not been defined
- Determine if mentor is a fit

Stage 2
- You have officially asked the individual to be your mentor
- You begin having 1:1 sessions
- You start to build a deeper level of trust and respect

Stage 3
- You and your mentor's relationship and trust deepens
- More effective coaching and 1:1 sessions

Figure 6.3 – The stages of mentorship

In this section, we discussed the three stages of a mentor-mentee relationship and what to expect at each stage. You learned that you will not gain the full benefits of mentorship until you fully trust and respect one another.

In the next section, we will discuss how to choose a mentor and the numerous factors that you need to consider when making your decision.

Choosing the correct mentor

Finding a mentor is challenging, but choosing the correct mentor is more challenging. A successful mentor will understand your goals, be someone you respect and look up to, understand your current situation and abilities, and see potential in you. We will separate this section into two parts: the criteria for choosing a mentor and asking someone to be your mentor.

Questions to ask when looking for a mentor

Let's take a visual look at the criteria I have found to be the basis for finding a good mentor:

Is the individual an **enabler** of my short- and long-term **goals?**

Is the individual someone I **respect** and look up to?

Is the individual **aware** of my **goals** and **current capabilities?**

Figure 6.4 – Questions to help you determine whether you are choosing the correct mentor

Is the individual an enabler of my short- and long-term goals? As we approach finding the answer to this question, we uncover another question directing the focus on your goals, *what are my long- and short-term goals?* If you do not have this documented somewhere, don't worry, we will cover it now.

Activity: Writing Your Goals

Grab any sheet of paper and split the page into four quadrants. Then, label each quadrant 3 months, 6 months, 1 year, and 5 years.

Next, grab some sticky notes and start adding your goals. Place the notes in one of the four quadrants.

Once you have this done, you can create a final copy in a Google Doc or any other type of file.

If you are comfortable, I recommend sharing this with your professional network or a subset of your network. We will come back to this point later.

Now that you have your goals, look at the goals that are in your one-year and five-year quadrants. As you start identifying potential candidates, ask yourself what skills and expertise the individual has that will help develop the skills that are necessary to achieve your goals.

Is the individual aware of my current goals and capabilities? This question is meant to access an individual's awareness of your current circumstances. It is more effective to have a mentor who has worked with you or interacted with you either virtually or in person at some point in the past. This is not to say you should disqualify anyone you have not had previous interactions with.

> **Pro Tip**
>
> If you have not had any previous interactions with an individual and still decide they are still a good fit as your mentor, you must be upfront with the additional time commitment involved in getting to know one another. This is a requirement for a fruitful mentor-mentee relationship.

If you post your goals on LinkedIn, individuals should be aware of your goals if they follow you. In the following section, we will cover another way to present your goals to your mentor.

Is the individual someone I respect and look up to too? This question contains two parts; both are critical if you wish to have a strong relationship with your mentor. You should eliminate anyone from your list who does not meet these criteria. It is not possible to have a mentor-mentee relationship with someone who you do not respect or someone you do not look up to.

Use the following Venn diagram to determine whether someone should be considered as a potential mentor:

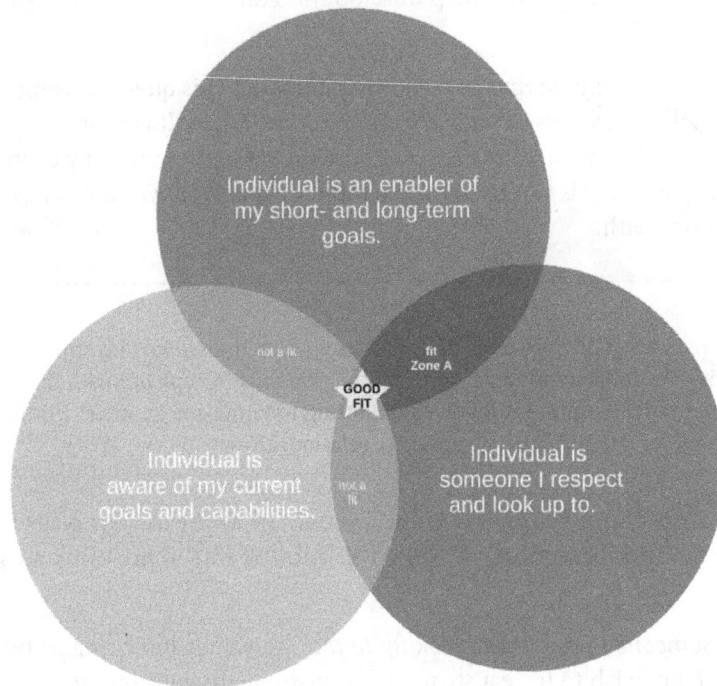

Figure 6.5 – A Venn diagram to decide whether someone is a potential candidate to be a mentor

In the next section, we will discuss the best way to officially ask an individual, who you have determined to be a good fit, to be your mentor.

Asking an individual to be your mentor

If you are fortunate, asking your mentor will be a natural progression in your already existing relationship. If you work closely with the individual and you are already having regular career-related discussions, asking them to be a mentor should be easy. This is the candidate who lands in the center of the preceding Venn diagram, as shown in *Figure 7.5*.

Pro Tip

There is a difference between an internal sponsor and a mentor. An internal sponsor is someone in your current organization who will help you with internal career progression. Interaction with a sponsor will likely end once you leave the organization that they are part of. A mentor will not be someone who will give you a job or get you a job; however, they will be part of your career journey in the long term.

Often, having both a mentor and a sponsor is beneficial. A sponsor will give you advice and career opportunities that are relevant to your current employer. A mentor will give you more holistic advice that is unbiased to a specific employer. *Both* are good opinions to have; especially when your mentor's advice aligns with a sponsor's job opportunity!

Most will not be so fortunate, and you will need to work a bit harder and move outside your comfort zone to land a great mentor. The following are several scenarios that you might find yourself in:

Scenario 1: A potential mentor is someone who you have worked with at a previous company, and you have stayed connected with on LinkedIn. This individual lands either in the center or in **Zone A** on the Venn diagram.

In this case, you could send them an email. Here is one example:

Subject: Mentorship

Hi [insert person's name],

During our time working together at [company] I was able to learn a lot and grow as an engineer. I have attached my career goals and feel with you as my mentor, we could further refine my goals and define a plan to help me achieve them. If being my mentor is something you would be open to, I would enjoy setting up an initial call to discuss.

Regards,

[Your name]

Scenario 2: A potential mentor is someone you met at a conference and have since been interacting with via LinkedIn. This individual lands in **Zone A** on the Venn diagram.

In this case, you could send them a LinkedIn message. The following is one example:

Subject: Mentorship

Hi [insert person's name],

Since hearing you speak on [topic] at [conference or event], I have been following you on LinkedIn and respect your understanding of DevOps. I have come to a point in my career where I believe I need a mentor to achieve my goals. I believe with your support and mentorship, I could grow professionally and achieve my goals. If you are open to mentoring me, I would enjoy finding some time on your calendar to discuss. Initially, I believe we would need to discuss time commitments.

Regards,

[your name]

Scenario 3: A potential mentor is someone you have never met but have been following on LinkedIn. This individual lands in **Zone A** on the Venn diagram; the likelihood of getting a positive response is not high because you have never met or interacted with this individual.

In this scenario, getting to know the individual on a more personal level would be the best approach. A person is far more likely to accept an invitation to be a mentor for someone they know personally.

In this section, we covered the questions that you need to ask to find a good, potential, mentor. Additionally, you learned about the potential ways in which to request an individual to be your mentor. In the next section, we will discuss when it is appropriate to request a mentor to be a reference.

Additional ways to get connected with a mentor

It is not always a straightforward or easy task to find a mentor. You might have a tough time finding a mentor within your organization or even within your network of connections. This is more common than you think, and there is an entire business built around connecting individuals with mentors to help them succeed. Should you be interested in additional support in finding a mentor, check out the sites listed here:

MentorCruise

MentorCruise (`https://mentorcruise.com/`) is designed to connect technology professionals: both mentors and mentees. Prices range from $150–$250 US per session. Services provided include a Q&A, interview prep, and resume review.

GrowthMentor

GrowthMentor (`https://www.growthmentor.com/`) is priced at 50 US dollars per month for a basic subscription. You can use the site to query a database of mentors based on your needs.

Pelion

Pelion (`https://pelion.app/`) is aimed at software engineering professionals, which includes those in DevOps. The rates start at $300 US.

Mentors as a reference

You have a mentor, and are beginning to apply for new roles, but is it appropriate to use your mentor as a reference? The answer is ambiguous, along with most of the things in this section, and should be based on your best judgment. There are a few questions that you can ask to make your decision easier.

The following is a decision chart that will help you simplify your decision:

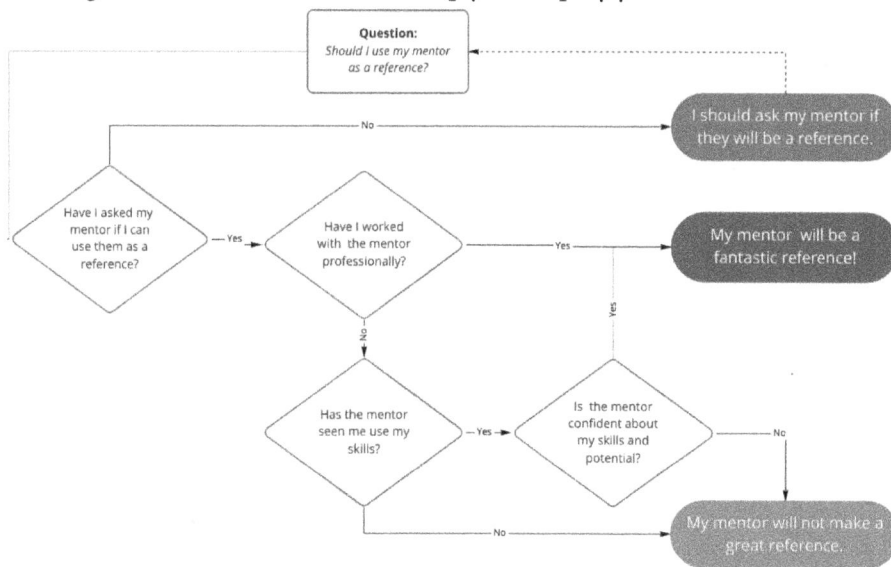

Figure 6.6 – The decision tree for "should I use a mentor as a reference?"

In the following portions of this section, we will cover each of the four questions, starting with the most basic one: have you asked your mentor if you can use them as a reference?

Have you asked your mentor to be a reference?

If you have not asked someone whether they are your mentor or otherwise, you can only use someone as a reference after you have asked them, and they have agreed. Using someone as a reference without their permission is unprofessional and can damage or ruin your relationship with your mentor. A mentor is going to be candid and honest with you. If they do not feel you are ready for a specific role, they would be placed in a tough position if they were asked to refer you for the job. If you ask them, this could lead to an amazing discussion and opportunity to grow from your mentor's insight. Although there is a chance they will decline, the odds are in your favor that they will gladly be your reference.

After a mentor has agreed to be a mentor, you need to ask a second question: have you worked with your mentor in the past?

Have you worked with the mentor?

If you have worked with your mentor, they will be aware of your skills. Also, the fact that they agreed to be a reference for the specific role shows they are confident you have in what it takes!

> **Pro Tip**
> If you have asked your mentor to be a reference for a broad job search instead of a specific job, you could ask the mentor about the specific role. In this way, they are aware of the specific role, and it could open up some great conversations on ways to better prepare for your upcoming interview.

If you have not worked with your mentor, you must ask another couple of questions to determine whether you should or should not use them as a reference. The first question you need to ask is whether your mentor has seen you use your skills.

Has your mentor seen you use your skills? Are they confident about your skills required for the role?

Seeing your skills and abilities is important for a referee to give an accurate representation of your skills, abilities, and character in their reference. They could have participated in a pair programming exercise or were present at a presentation that you gave at a conference. If the individual has not seen your skills, personally, I would not use them as a reference; you should use your best judgment. If you are uncertain, have a conversation with your mentor.

Next, you need to ask this question: is my mentor confident in the skills I require for the role? If you asked your mentor to be a reference for a specific role and they have agreed, you can assume they are confident in your abilities for the role and use them as a reference. On the other hand, if you asked your mentor whether you could use them as a reference in general, not for a specific role, you will need to access their confidence in your abilities for the role. The best option would be to discuss the specific role with your mentor and get their feedback on it.

In this section, we covered the key questions you need to ask when deciding whether you should use a mentor as a reference. The most effective and sure way of determining whether to use your mentor as a reference is by discussing the specific role you are applying for.

Summary

In this chapter, several topics on mentorship were covered, including the importance of mentorship, the relationship dynamics between a mentor and a mentee, how to find a mentor, and using a mentor as a reference. In *The importance of mentorship* section, we covered assistance and guidance on setting achievable goals, career coaching, motivation, and career advice.

In *The mentor-mentee relationship dynamic* section, you learned how there are three stages of a mentor-mentee relationship. Stage 1 is where you have not officially asked someone to be your mentor, stage 2 is where you officially define the relationship and begin having official 1:1 sessions, and in stage 3, your relationship grows due to an increase in trust and respect for one another. It is at stage 3 where you are able to begin having more productive and fruitful discussions.

Next, we discussed how to find and choose a mentor. In this section, we discussed three questions that you should ask to determine whether someone will make a good mentor. The three questions included *Is the individual an enabler of my short- and long-term goals?*, *Is the individual aware of my current goals and capabilities?*, and *Is the individual someone I respect and look up to?*

Finally, we discussed whether you should use a mentor as a reference; the key takeaway from this section is that you can always have a discussion with your mentor. However, if time does not allow this, there is a decision chart that can help you.

In the next chapter, we will discuss how working with recruiters can increase your chances of landing an amazing DevOps job.

7
Working with Recruiters

Navigating the recruitment process is a form of art and will remain a mystery until you go through it a few times. It is important to understand this as you go into this chapter, especially if you are preparing for your first interviews. Don't get discouraged if all the information does not make sense at first; some will not until you have experienced it first-hand. This chapter will focus on the importance of relationships with recruiters and navigating those relationships with diverse types of recruiters at various stages of the interview process. We will dive deeper into the intricacies of the interview process in *Chapter 8*, *Preparing for Your Interview*.

It is difficult to find a job without the help of recruiters. In this chapter, you will learn how to build lasting relationships with recruiters and use those relationships to land a job.

The following topics will be covered in this chapter:

- Different types of recruiters
- Where to find them and how they can find you
- How to present yourself
- How to negotiate
- Following up, but when?

Different types of recruiters

There are several types of recruiters. While we will not provide a comprehensive review, we will attempt to cover the most common scenarios that we have encountered throughout our careers.

The following diagram provides an overview of the three types of recruiters you may come across while looking for a job:

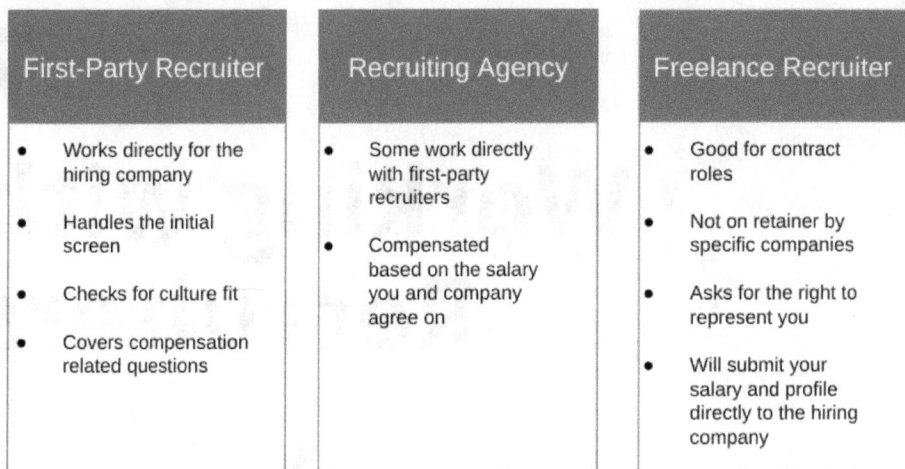

First-Party Recruiter	Recruiting Agency	Freelance Recruiter
• Works directly for the hiring company • Handles the initial screen • Checks for culture fit • Covers compensation related questions	• Some work directly with first-party recruiters • Compensated based on the salary you and company agree on	• Good for contract roles • Not on retainer by specific companies • Asks for the right to represent you • Will submit your salary and profile directly to the hiring company

Figure 7.1 – Overview of recruiters

First, let's take a deep dive into **first-party recruiters**.

First-party recruiters

This is a recruiter that works directly for the company and is trying to fill a role. They may reach out to you through a social network, most notably LinkedIn, or another job board. If you apply directly to the job on their website, this is the person you might have your first call with. Some companies even divide this into separate sub-roles, where one is the source of leads, and another does the actual screening. Regardless of whether you meet their criteria, you get access to the interview loop and the team.

They are also generally who you would speak to about your salary requirements, and ask questions regarding compensation, bonuses, and other areas that may interest you. They, in turn, try to see how well you align with the job description.

While this is not a technical interview, don't get too relaxed! Sometimes, recruiters or HR personnel screen for culture fit and not just skill sets. This means they may be paying attention to several things, including how polite you are, how much you know about the company, and if the screen is on video, your tidiness, or your persona.

Make sure you treat these calls as part of the interview but don't be afraid to ask questions and make sure this is a role worth your time. You don't want to be one of those people that finishes a six-round interview process only to realize there is some downside that could have been foreseen with the right question.

Recruiting agencies

A recruiting agency falls into one of two categories – an independent company or one that's retained by the end client. An independent company may find your profile and submit it to several different jobs. A company on a retainer is searching for talent that specifically matches a company profile, but if they are large enough, they may have more than one viable candidate company.

Agencies tend to make either a bonus based on your salary or a percentage of your income (or a fixed amount) for a set period.

Generally speaking, the more you make, the better they do, except in situations where there is a lot of competition for the role(s) and agencies try to undercut one another. Try to develop a sense of whether this is the case so that you know how to negotiate appropriately.

Freelance recruiters

A freelance recruiter usually approaches you on LinkedIn or through an unsolicited email. They will pass you their job requirements and try to get on the phone with you as soon as possible. Most of the time, they are not final clients and may refer you to an agency or multiple recruiters before finally talking to the company. While they may have insight into roles you do not know about, they are typically not on a retainer and the roles are freely available. It is debatable whether it is better to apply directly than to apply through a recruiter. It depends on whether the recruiter already has a relationship with the hiring manager.

In my experience, freelance recruiters are the best for contract jobs, where you specify, say, your rate, and then the recruiter can figure out whether they can make it work for that amount.

They ask for the right to represent you via an agreement sent in an email, although sometimes, this is more formally in a signed PDF, and want to confirm the rate or salary that you agree to be submitted under in writing.

The recruiter's role at various stages of the interview process

So far, we have discussed three types of recruiters. In this section, we will look at where the three types of recruiters may be involved in the various stages of the interview process. These stages, as shown in the following table, will be described in more detail in *Chapter 8, Preparing for Your Interview*:

	Talent Discovery	Initial Interview	First Round	Technical and Follow-Up Rounds
First Party	Not usually until the resume is submitted to internal system.	Highly likely, unless the company has employed the help of an agency to perform this initial screen	Almost always as this is usually with the hiring manager.	Sometimes, the technical round is contracted out, but the follow-up round coordination is usually done through the first-party recruiter.
Agency	Will reach out directly via social platforms, or after submitting your resume to the internal system.	Not unless a company has contracted the agency to perform this step.	Not a common practice.	Sometimes it is common for the agency to take on the technical interview, but they will rarely handle the follow-up coordination.
Freelance	Yes – freelance recruiters rely heavily on social platforms to find talent.	Only for direct-hire positions; common with contract roles.	Not a common practice.	No, except for specific contract positions.

Figure 7.2 – The recruiter's role in various stages of the interview process

Talent discovery – that is, actively pursuing social platforms for valid candidates – is not usually something first-party recruiters take part in; they usually post open roles on social platforms and wait to reach out to candidates until they apply through their talent management system. Agencies are contracted for their ability to go and search for talent; an exceptionally sizable portion of candidates that agencies find are from social platforms or past relationships they have built. A freelance recruiter's only source of income is finding talent and presenting that talent to companies. They are considered the best headhunters because of their unnatural ability to find talent where others miss it. The reason they were able to become a freelance recruiter is because of the reputation they have built up with both candidates in a specific job market as well as employers.

The initial interview is usually taken care of by the first-party **human resources** (**HR**) recruiter. The exception to this is when a company has contracted an agency to cover this portion of the interview process or for certain direct hire or contract roles.

The first-round interviews are scheduled through the first-party recruiter and are usually with the hiring managers. The exception to this rule is certain contract positions in which a manager has a good relationship with either an individual freelance recruiter or agency and trusts them to take care of this step.

Technical interviews can be conducted both by an internal first-party process or through a recruiting agency. It is not common for freelance recruiters to take this step, though there are exceptions where this may occur. Follow-up rounds are usually scheduled with internal resources, such as team members or individuals from other parts of the company you will be working with, and will be coordinated by the first-party recruiter.

Where to find them and how they can find you

In a general sense, if you maintain an active social persona and have an updated resume, recruiters will always find you first. There are many job boards, and some recruiters buy entire databases of candidates so that once you are on the job board, you may continue to receive emails even after you are no longer looking.

With LinkedIn, attracting the best recruiters is a simple affair. Make your profile and resume look the same (roughly), with similar summaries and keywords. Make these keywords match your experience and the keywords that are used the most in search terms and the jobs that interest you.

You can also add or message recruiters – first- or third -party – directly about the jobs they have posted on. While I do not recommend filling your LinkedIn with recruiters only, I do think it is normal for 1/3 of your contacts to be related to job hunting or career growth. That is the primary purpose of LinkedIn.

> **Pro Tip**
>
> Recruiters search too, and when they search for *Terraform*, it pays to have that on your profile and resume. Do not add skills that you do not have, as you may be asked for or even tested on them later. However, do scan different job posts for these keywords and use them as a guideline of what skills to learn or develop further.

Finally, making sure your profile is visible and states that you are open to work or open to the conversation will tell the recruiter that it is OK to contact you. A good relationship with a recruiter can yield you multiple leads throughout your career. A great profile on LinkedIn can keep you rich in contacts and ensure you never miss any career-enhancing opportunity!

How to present yourself

I have briefly touched on some aspects of presentation and how to make yourself appealing to recruiters. Let me reiterate everything I think is key.

The first part of highlighting how you present yourself is beefing up your LinkedIn profile. We covered this in great detail in *Chapter 5, Building Your Network*, but I will say that having a professional page that has a high-quality picture and high-quality content is key. What makes content high quality? Part of it is the writing and part of it is the content. In general, I recommend having your profile and resume be the same or very closely related. I create my resume based on my LinkedIn profile, and whenever there is an update, I do it on the profile first, then export the resume later.

Having someone proofread your position summaries and making sure they are descriptive enough goes a long way to having a recruiter read them. In a technical field, you must describe the technical skills you used for each job. Rather than speaking in generalities, be specific. As an added tip, make sure you include numbers and percentages when highlighting results.

Stressing the most important (and marketable) skills in your summaries is key not just for the reader but for the search engine that is looking for those skills so that it will return the best matches.

Once you feel like your profile is professional, descriptive, and comprehensive when it comes to all the areas mentioned, it is time to state your goal of finding a new position. LinkedIn has different tools to accomplish this, including having custom badges that state you are looking for work. For the more privacy-conscious, you can make it so that only recruiters see that you are open to new opportunities. This signal will let recruiters know to contact you. With a good enough profile, you should have a steady swarm of messages and contact requests. I wouldn't add every recruiter to your network as you want to keep it balanced with people in your field and with people that will improve your domain knowledge or some other aspect of your career. You can determine your ratios for yourself, but just know that it is not necessary to connect with a recruiter to apply for one of their roles. If there is someone you enjoyed working with or that has particularly good roles, then it pays to network with them.

So, you start getting messages – then what? A lot of the introduction messages will provide a brief version of the job description. Some will contain all the information you need to determine whether you are interested, while others will contain nothing at all except a message expressing a desire to connect or have a quick phone call. My advice is to do some casual probing and state your requirements before getting on a phone call. Otherwise, you will spend a lot of time sharing information about yourself only to discover the role was not for you.

I would request at least a job description and a salary range (annual or hourly) at a minimum. Some may not give it, but in my experience, most will, especially if you state that you do not want to waste their time. We will cover this in more detail in the *How to negotiate* section, but you want to convey what you are looking for and what you are not, and only make appointments or have calls with those opportunities that more closely align with what you are looking for.

Something else you can do to improve how you are perceived is increase your social awareness and build your brand. We covered this in great detail in *Chapter 5, Building Your Network*, but gaining and displaying certifications, authoring articles, or reposting other articles in your feed and even authoring books (like this!) can increase your name recognition and brand, which can make you more appealing to recruiters and prospective employers.

Finally, I will say that you can be proactive about connecting and messaging with recruiters and hiring managers. LinkedIn has a professional version that allows you to send messages to people outside your network and that you can use to query their database. It becomes a little bit cold calling in sales at that point. I prefer to build a great profile and let the recruiters come to me.

How to negotiate

The way to negotiate depends on which part of the job hunt you are on. When you are dealing with recruiters, you are mostly establishing your rates and your desired wages. Other things you may want to establish are your expectations for bonuses, stocks **restricted stock units** (**RSUs**) or options, for example, **paid time off** (**PTO**), and whether you expect to work from home or work at the office. Increasingly, people (especially in tech) desire the flexibility to work remotely at least some of the time because they can do their job if they have a computer and an internet connection. Employers have different expectations, so it is important to check with your point of contact to see what their expectation is.

So, what are RSUs? According to investopedia.com, an RSU refers to a form of compensation that's issued to an employee in the form of company shares. What makes them special is that they have a vesting schedule, so you will typically earn a percentage after a certain amount of time – say, 25% per year. However, RSUs can also be tied to performance metrics, not just employment longevity.

What if you don't know how much you should be asking for? Well, you can ask some jobs what they pay for and use that as a baseline. Not everyone will offer this information, but some will, and that will get you started. You can raise your asking price until you get resistance, or you can leave things fluid and ask for a range. As a rule of thumb, the lower end of your ask is what will be used to make sure that is a number you are comfortable with. If you get more, great, but always ask for what you need straight out of the gate.

Things that do not work

Some mistakes I have made are valuing cash over the job's title, valuing cash or title over culture or fit, or going in at a lower wage while thinking I could work my way up. Sure, it could happen, but we want to deal with the more realistic scenarios.

> **Pro Tip**
> Find out how much someone with your experience makes for a similar role and then ask for that.

While your internal voice may want to deter you from asking for much more than you are currently making, consider the possibility that you are underpaid, or that you have been working in a local market and now have access to a global market thanks to remote work. If you don't ask, you won't know. The worst thing they can say is no.

In my experience, hope is not a strategy. Either you try to get what you want upfront, and not get it exactly but approximate as much as you can, or you are better off continuing to look.

I remember when I was looking for a leadership role and I saw a Director of DevOps role that seemed attractive. It was with a famous firm, so there was some prestige associated with the role, and from my initial conversation with the recruiter, the compensation seemed to be in line with what I was looking for. I did not ask enough questions and deep within the job description, there were more details that I could have used to ask sensible questions and make sure that the role was right for me.

For example, I could have asked how many people would this person have reporting to them. I was hoping to lead more than one team as I had several years of experience leading 5-10-people teams and thought I was ready for more. I had noticed that for a lot of senior roles at big companies, managing managers was a requirement. I had led senior folks but never managers, so this was something I was hoping to get in my next role.

I did not ask, and when I had my interview with the hiring manager, they mentioned that the role was primarily for a technology leader and that it would only have two direct reports. Hardly a big team for a director! Some companies treat directors as something functional (you lead a group, so you are a director), while others treat it as a rank (from a senior manager, you get promoted to a director!) Some, however, band the roles based on compensation, and functionally, it could be quite different from the same job at another company with similar pay.

What did me in was something else. The manager was looking for someone with deep experience in one piece of technology. I had some experience, but this was not my specialty, and the hiring manager made it known that this was his priority. I was frank (as I tend to be) and said that my experience was with a competing product, but I was confident in my ability to bridge the gap very quickly. He mentioned that he went through the same process and that it took him a long time to get caught up, so he assumed it would take me a long time as well. That was nonsense, but I was not interested in changing his mind. The lack of a team made the role unattractive, and I left the interview feeling like I had wasted everyone's time. Read and ask questions first!

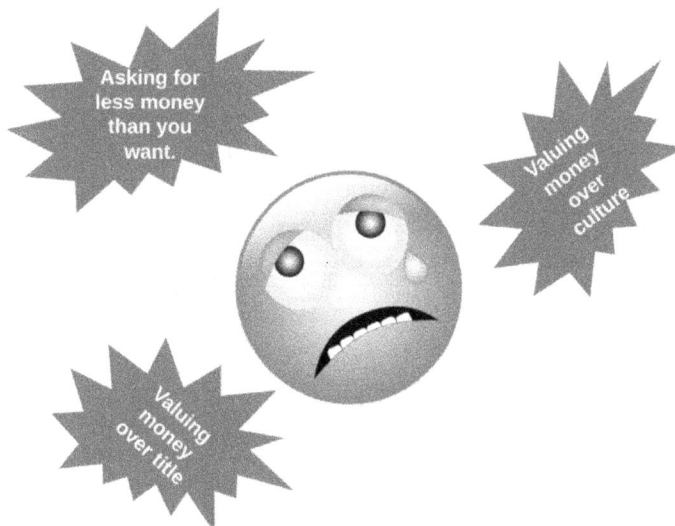

Figure 7.3 – Things not to do when negotiating

Next, we will discuss things that do work when you're negotiating.

Things that do work

The most important thing when it comes to negotiation is doing your research. Ask other recruiters, friends, colleagues, and even those on social media. Establish a far baseline and communicate early. You don't want to waste your time on opportunities you would not take.

Find out about other rewards beyond compensation. It may affect how satisfied you are with your job and some companies do things differently. You cannot put a price on a good culture, so try to get a sense throughout the interview process of whether this job is a good fit. Ask yourself whether you see yourself there 5 years from now. Do the people that interview sound like people you may want to hang out with? You cannot infer all this from a quick call with a recruiter, but you can certainly ask them about their culture and establish some baselines.

One thing I discovered from personal experience is that salary adjustments are much slower when you stay in the same company year over year. You may get a 2-5% raise, but changing jobs may get you 15-20%. Doing your research allows you to determine the salary based on current market conditions, not just based on your career.

One thing to remember is that salaries used to be tailored toward your living location. Remote work and competition have lessened this a bit, but there are still companies out there that penalize you for living in a lower cost of living location. Having a high wage while working at a lower cost of living is called arbitrage, and it is something you can take advantage of while working in the tech industry.

Working remotely is another huge factor in cost reduction. You can eat at home, save on gas, parking, and general car maintenance, and partake in the salary arbitrage. I always treat working remotely or remote benefits (working from home 20% of the time, for example) as part of the negotiation process. It has made a significant difference to my quality of life, and I consider it as important as asking about benefits or culture. Conversely, you do not want to stay silent and find out, after spending weeks interviewing, that this amazing job requires you to drive 45 minutes each way or relocate!

Now, there is nothing wrong with relocation, especially early on in your career. Getting out of your hometown and into a place where tech jobs are aplenty might be advantageous, and a company can certainly help you with the moving costs. Unfortunately, arbitrage works both ways and you can end up getting a raise but having the cost of living be so much higher that you lose whatever benefit you gained. You may even end up on the negative if you move to a much more expensive city! This is dangerous in the tech community because cities such as San Francisco, Seattle, New York, and LA have a lot of jobs, but they are also some of the most expensive cities in the country! Even places such as Austin are much more expensive now than they once were, so you want to research the cost of housing (more than other metrics) when comparing cities.

The best tip is to try to normalize your salary and adjust it for each city, using a website or calculator. That way, you can ask for a raise and any adjustments needed for cost-of-living differences and know exactly how much to ask for. As usual, data is king, and you can always use hard numbers and research to back up your claims. Recruiters will certainly say that they will pay a salary commensurate with experience, but they may be looking at the nationwide numbers, not a number that's been adjusted for the specific location being considered.

If you look at a list that specifies the highest cost of living, you will find that, for example, San Jose is in the top 10 most expensive, so even if they pay high wages, your lifestyle might be lower than if you live in Raleigh, NC, for example. Such data is available in the US from the Bureau of Labor Statistics.

Besides a base salary, you may want to negotiate in other areas of compensation, such as bonuses and equity/stocks, which can be awarded as options. Here, you have the option to buy tomorrow at an earlier price, or with RSUs, which are stock awards with a multiyear vesting schedule. Usually, you get a yearly amount, and they take 4 years to vest, so every year, you are simultaneously earning and vesting more. This is common in big tech companies (such as **Meta**, **Apple**, **Amazon**, **Netflix,** and **Google**) and when you get high enough on the career ladder and become a senior leader. For some roles, the equity component can be worth more than the salary! It depends on where you are in your career and your company.

> **Pro Tip**
>
> Sometimes, companies will say a bonus is a certain percentage but will not explain that that bonus might be dependent on something else, such as your performance, company performance, or other factors. It could also be discretionary and not be awarded at all! Some companies tend to band the bonus payout to performance bands, so if you get a 4 out of 5, you may receive 80% of your bonus. Finally, some companies pay a percentage of your bonus after 1 year, and the remainder after the next, incentivizing you to stay with the company.

History shows that if you work in a successful company, have stock awards, and stay while the company grows, you can potentially make a lot of money as the stock of the company grows every year! Now, you must be mindful because not every company grows as much as Netflix or Amazon do, and some companies tank and go under.

When you are dealing with start-ups and taking equity as part of your compensation, consider this a bonus and not a core component of your pay. After all, it could be years before the company has a successful exit and you may never perceive any value from that equity package. From that perspective, established public companies have more leverage when offering stock awards.

Figure 7.4 – Things to do when negotiating

In this section, we learned how to negotiate when working with recruiters. Things that work, as well as things to try and avoid, were discussed.

In the next section, we will discuss how and when to follow up after applying for a role.

Following up, but when?

The most stressful stage in a job search is waiting for a call, email, text, or smoke signal after submitting your resume for a job. This section will be broken down into two parts:

- The waiting game
- Etiquette for following up with recruiters

Let's start by covering the waiting game.

The waiting game

After submitting your resume, it is common to want feedback instantly; I have been there many times and know the debilitating feeling of not knowing what's next. Be patient – a general rule is to wait at least 2 weeks before following up. Many factors affect the time it takes for your application to be processed by the hiring team:

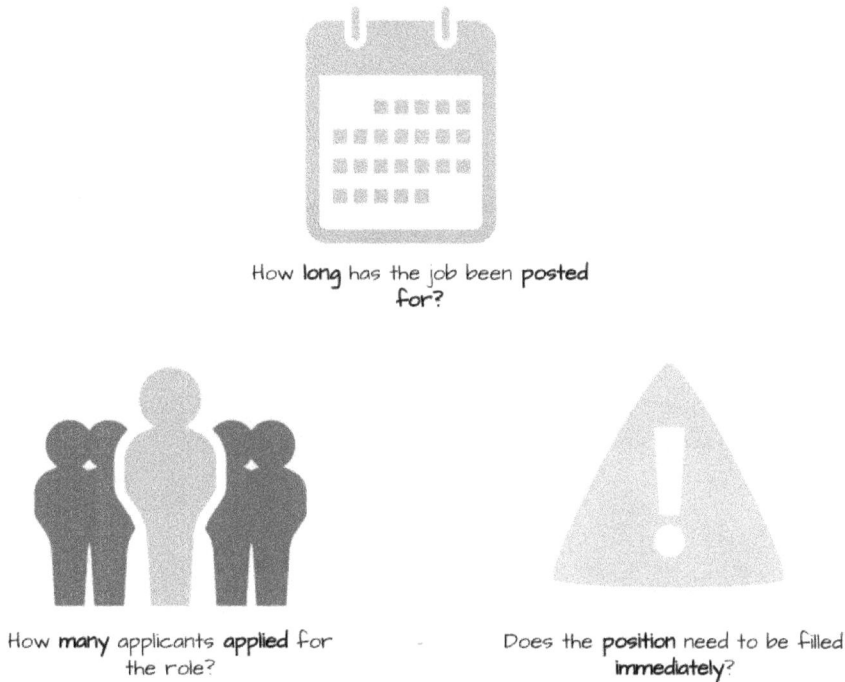

Figure 7.5 – Reasons for the time delay in hearing back from recruiters

How **long** a role has been posted affects the timeframe you should expect to hear back within after applying for a role. The longer a role has been posted, the more likely you will get a response quickly. The time when a role was posted is not overly telling of an expected response but when it's combined with the number of applications that have been received for a role, you can get a good estimate.

The number of applicants a role has is a multi-faceted metric; in this section, we will only consider how it affects the response time from a recruiter. The more applicants a role has, the longer you can expect to wait. In a situation where a role has been posted for a long time and has many applicants, your wait will be either exceptionally long or quite short. There is a possibility that the reason the role has been open for so long is that no suitable candidate has been found.

The urgency to fill a role is something you can use to your benefit. You will hear back quickly from the recruiter and will also have an expedited interview process. However, you must ask why there is such an urgent need to fill the role. In my experience, as both an interviewer and a candidate, trying to fill a role overly quickly is in no one's best interest. An expedited interview process may be a result of toxic company culture or poor planning, and it may result in an inability for both parties to get an informative read on whether the relationship will be fruitful. On the other hand, you will never know if you don't try.

At the end of the day, waiting is not a lot of fun but understanding the factors that affect the wait time can make it more bearable.

In the next section, we will discuss the best follow-up practices to follow when working with first-party recruiters.

Etiquette for following up with recruiters

A few weeks have passed since you applied for a role and you have heard nothing; before you send an emotion-fueled email – or worse, call them – take a moment and try to follow some best practices. These can be seen in the following diagram:

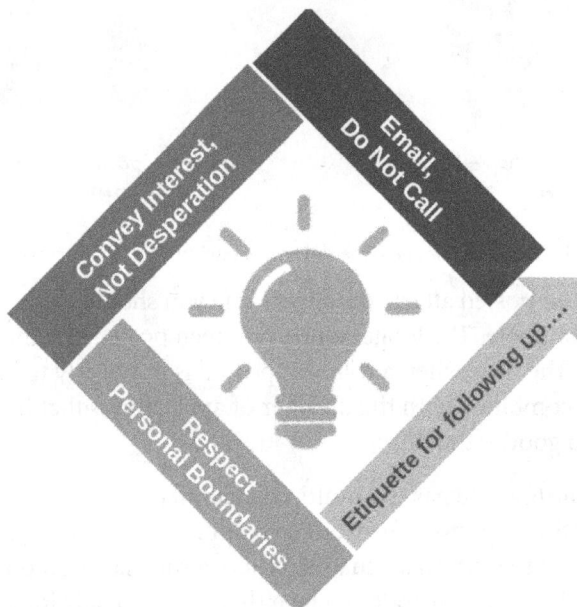

Figure 7.6 – Etiquette for following up with recruiters

Email is the only acceptable method when you're following up with recruiters. Calling is not appropriate unless you have been given permission ahead of time. Email provides digital documentation of your correspondence as well. Some companies, especially those who receive a large volume of applications, ask applicants not to follow up; you should adhere to the company's policies.

Show interest while avoiding desperation as you compose your email. Reiterate your sincere interest in the role and include specific examples from the job description; as shown in the following screenshot:

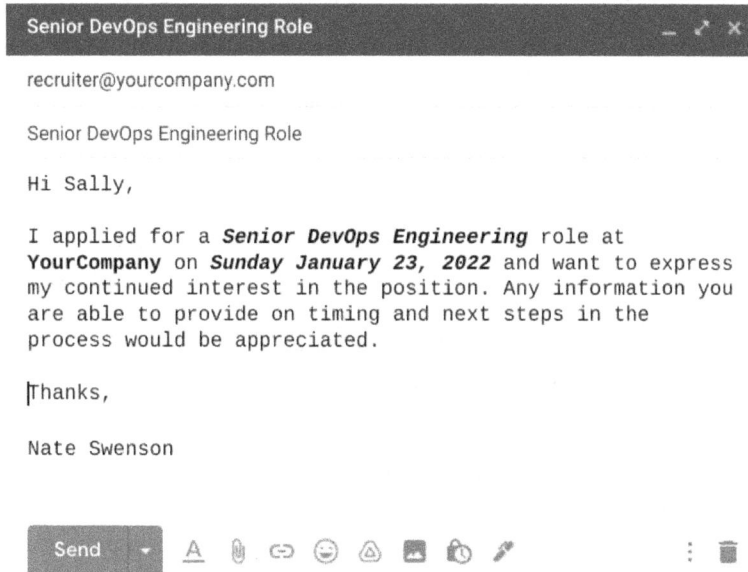

Figure 7.7 – Email example

Respecting personal boundaries is as simple as being professional. Recruiters are not looking for friends – they are on the hunt for top talent. First, I advise against adding recruiters to any non-professional social media such as Facebook. First-party recruiters work specifically for a company, not for you, which means they owe you nothing, not a response or follow-up, no matter how much you want one. For example, if you followed up more than once and they didn't get back to you, then they are just not that interested, and it is time to move on.

Keep in mind that it is better to keep a bridge open and in decent shape versus burning it by being unprofessional. Circumstances change and down the road, that bridge may land you your best job yet!

Summary

In this chapter, you learned the basic rules of dealing with recruiters. You also learned how and when to follow up and you read up on some of our experiences looking for jobs and negotiating with recruiters, hiring managers, and agencies.

Having a professional profile that shares your experience and carefully highlights all your skills is supremely important. Your online profile is your calling card, so having the correct set of skills and experience can get you the type of contacts you want both from a networking perspective and for specific opportunities.

You can be proactive with your search, and you can make it so that you can receive daily job hits based on your online profile. Make it a habit to quickly classify each job and put it in the right bucket. Contract gigs and salaried roles can vary wildly in responsibility and compensation, and you must make sure you know the answers you need before you start the interview process!

Do your research, ask relevant questions, and read the job description to make sure you know what you are signing up for. The internet makes it easier than ever to apply for jobs, but blindly doing so may cost you more time down the line. Find out about their salary ranges, share your expectations, and determine whether they are in line. Ask about benefits, culture, remote work, and other forms of compensation, such as bonuses, equity, or stock.

Be mindful of cost-of-living differences and realize that a dollar in San Francisco is worth less than a dollar in Boise. Ask companies about their relocation packages, salary adjustments, and ability to work remotely. Moving for a new job could prove to be a great career move, but you certainly don't want to do it at the expense of a lower quality of life.

As far as recruiters are concerned, be frank, be direct, and try to not waste their time or yours. If an opportunity seems like a terrible job, don't interview just for the sake of it. Sure, getting interview practice is valuable, but it's better to do it with jobs you are genuinely interested in.

Connect with recruiters who you enjoy working with, and feel free to network with people you meet in the interview process, including hiring managers. Some may not accept a request at that point because they are still in the interview process, but you may reach out after the fact and state you would like to connect regardless.

In the next chapter, we will begin discussing how to prepare for your interview.

Section 3: Interview Process

You did it! You landed an interview! Up to this point, you have been building an online façade of yourself. In this portion of the book, we will detail how to clearly showcase your talent to the hiring team.

This section comprises the following chapters:

- *Chapter 8, Preparing for Your Interview*
- *Chapter 9, Interviews Step by Step*

8
Preparing for Your Interview

You've talked to recruiters, gone over the job description and the level of remuneration, and set up an interview. This can be a daunting process, so we are going to spend the next several pages breaking it down, talking about all the steps involved, the best preparation practices, and other tricks of the trade. In the following chapter, we will go into more depth and even talk about non-typical interview cases.

The following topics will be covered in this chapter:

- Phases of the interview process
- Best ways to prepare
- What to expect
- Tricks of the trade

Phases of the interview process

Usually, after the initial setup call, your first stop is with the hiring manager. Obviously, this can vary from place to place, but I find it to be the most common starting point. In our field, you can expect a technical screening round, which can be conducted by an individual or by a panel. For lead or management jobs, this can be conducted by your peers or maybe even by the people that report to you. Depending on the company, this can lead to another individual call with someone from your future leadership, which could mean your future boss or even their boss! Things they look for here are culture fit, the ability to communicate, and general compatibility. We will go into detail later. In some places, they may have more rounds. I have seen design-focused interviews, coding challenges, situational questions, behavioral, and even out-of-the-box thinking. I will cover this in *Chapter 9, Interviews Step by Step*.

I also want to mention the fact that some technical jobs have programming tests, which could be offline or live, with someone watching you code and asking you questions. Some companies may also ask you to undertake behavioral or cognitive tests. I will cover these things in the non-typical interview walk-through. Coding tests might be common for software engineers, but much less so for DevOps professionals. I think three rounds is probably the most common setup. If you are being interviewed at larger companies, such as Fortune 100 companies, the process may entail four or five rounds.

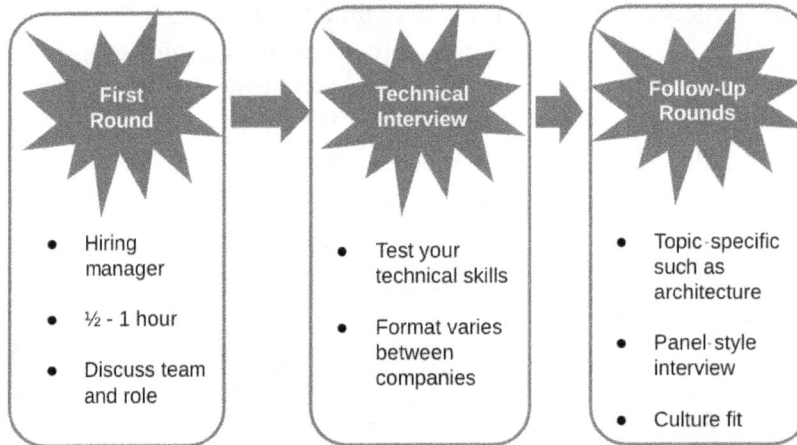

Figure 8.1 – Interview stages

For a lot of places though, it's three rounds, short and simple. Preparing for your first round is not difficult, but it's one of the places where a lot of candidates leave work on the table, so let's start there.

First-round interview

While you may speak with a recruiter and a person from human resources first, we consider the first-round interview to be with someone from your team. Typically, this is the hiring manager. This interview can last from 30 minutes to an hour, and it usually involves the hiring manager telling you about the company, the team, what they need, and then asking you for your experience. Make sure you have done your homework and you know something about the company.

If you have read the job description, you should know roughly what they are looking for and how your experience matches their search. When talking about your job experience, you should highlight your skills and anything you think is relevant to the position you are applying for.

You should start by providing an overview of your career and spend 5 minutes highlighting who you are and what you have been doing. If you can address how you and your experience fit into this role, all the better.

One question they will certainly have for you, especially if you are not fresh out of college, is why you are looking to work at their company, or why you are looking to make a change. Usually, you can say you are looking for better pay, but that will not convince your future employer that you will not leave them for better remuneration elsewhere in the future. It is better to tie your desire for a new role to your career development and to the company you are applying for directly. This is a fantastic opportunity for you to show enthusiasm and interest in working with them.

Something I have noticed that hiring managers do at this stage is trying to determine how you would fit into the team. For example, if you are a manager, they will ask you about your leadership style. They may also ask you situational questions, such as *you telling them about a time* when you dealt with a particular issue. They may also ask you how you see yourself in terms of strengths, weaknesses, and career prospects, such as where you see yourself in 5 years.

Depending on who you are being interviewed by, you can expect to answer a few or the majority of these questions. Amazon in particular places great stock on this type of interview question.

> **Pro Tip: First-Round Prep**
>
> As a general best practice, have a few stories pre-prepared about how you have dealt with common scenarios. For example, tell me about a time when you disagreed with a stakeholder and how you went about resolving this. People want to know you can handle disagreement and friction, and are resilient enough to survive the modern workplace, so have a few stories to hand that showcase how you dealt with, and overcame, adversity and similar situations. Keep a spreadsheet or document handy and practice, and this will become second nature.

Once they have told you about themselves and the team, and once you have gone over your background and answered whatever questions they might have, they will certainly ask you whether you have any questions. This is a great time to show that you did your homework and ask about the company, its future plans, and how you can contribute to making their efforts better.

Other important questions you can ask include the company's next steps and whether there are any reasons why you may have a challenge filling this role, or even better yet, what they think is necessary to be successful at this role.

Technical interview

This round can look quite different depending on where you interview. It could be a coding test, where someone gives you a problem and watches you solve it, possibly providing you with prompts followed by a period where you are required to work on the solution. During this time, they may ask questions regarding your work. It could be a panel interview, where three peers or experts ask you technical questions to see how deep your understanding extends across multiple domains. This is the most important round because you will be hired primarily for what you know, not for what you have done, and this is your opportunity to show it. Do not be afraid to ask questions if you need to, or to say *I don't know* when you are not sure. This is better than rambling, where you can end up digging a hole for yourself. You can always say *this is not my area of expertise* and prompt the interviewer to ask you a different question. Nobody knows everything, and so it is extremely easy to be stumped by someone who is trying to stump you. What you want to show are thoughtful answers and that you think about solutions in an intelligent and creative way.

Pro Tip: Technical Interview

The technical interview is the most stressful part of the interview process, and it is designed to be that way. Use this to your benefit, stay calm, and allow yourself to process what is being asked of you. Ask for another question if you are not familiar with the concept or language of the one given to you.

Usually, at tech giants (Google, Meta, Amazon, Microsoft, and so on), there will be a difficult technical assessment that challenges your knowledge of computer science, including data structures, critical thinking, objects, and Big O notation. For cloud-related jobs, you may be asked specific questions on cloud services akin to those asked in certification exams. It pays to ask the recruiter what type of interview you will be facing next and whether there is any preparation advice. When there is a coding challenge, typically they employ a platform that allows you to use multiple programming languages, so you can generally use the one you are most familiar with. I always use Python, on account of how simple it is to use, but if you are more comfortable with Java, go for it. The key is to be proficient with at least one programming language. For these types of interviews, it is best to get plenty of practice, as stated previously. The following URL, `https://leetcode.com/`, and *Cracking the Coding Interview* (a book by Gayle Laakmann McDowell) are great resources.

You are also expected to be familiar with Linux unless otherwise specified, and this implies that you are familiar with Bash and shell commands and scripting. Depending on the position and the company, you may also need to talk about architecture or specific systems.

Finally, there is the possibility of take-home assignments and homework. This can vary from architectural diagrams and design documents to coding challenges and actual services deployed in the cloud, which they will connect to and verify. If they are take-home assignments, time is on your side, so do not panic; just set aside a day when you can focus on this.

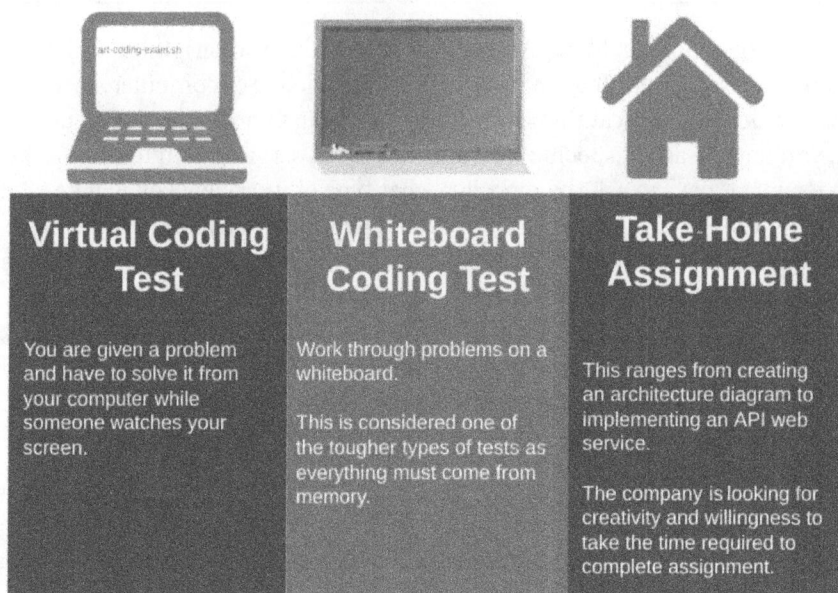

Figure 8.2 – Technical interview types

We will talk more about exceptions to the interview process in a future chapter, but right now, let's discuss the follow-up round.

Follow-up rounds

There could be one or more follow-up rounds, depending on the company and position. For example, there could be a dedicated architecture or design interview where an expert in that area poses a scenario and wants to see how you approach the problem. If you are in leadership, it could be meeting a fellow leader from another team, or even your boss' boss. Interviewing with other teams is not uncommon, especially when there is clear overlap, such as with QA, testing, and DevOps. If you are an individual contributor, you can meet a fellow engineer, a more senior engineer, or an engineer from another team that collaborates closely with yours. DevOps is very collaborative, so you will meet people outside your own team, and it is important to show that you are customer service-oriented and a team player.

If you are in leadership, you can expect a panel interview for your technical knowledge and another panel interview for your leadership and collaborative qualities. These tend to be with other leaders in the org, usually peers or individuals at a similar level.

It is also common to meet with the person above the hiring manager, as you are part of their org. This will not be a technical interview, but a combination of a culture fit, leadership style, and aspirational qualities check. If you are a leader, be prepared to speak about your leadership style in depth, with concrete examples. If you are an individual contributor, you may be asked a growth-related question, where you are expected to share career plans so they can see how your plans align with their needs. It is common to be asked where you see yourself in 5 years, so having a rough outline in mind is advantageous.

You should hear a decision within a week of your last interview, but this can fluctuate if they are interviewing many candidates. Always follow up with the recruiter to find out the next steps but give them some time as even consolidating all the feedback from multiple interviews can take time.

Best ways to prepare

Before interviewing, the best way to prepare is by researching the company, your hiring manager, and the position. The job description document will have a lot of information regarding what their ideal candidate looks like. If you lack a skill that is highlighted there, you could try beefing up your knowledge in that area. If there is something required in which you are an expert, make sure you highlight this.

Pro Tip: Technical Round Prep

For the technical round, it's best to be prepared to have your résumé challenged. Everything in it is fair game, so don't misrepresent yourself!

Ask the person responsible for scheduling your interviews or your contact in HR what to expect in each round, and doubly so for technical rounds. You can get sample questions or useful feedback from prior interviews if you just ask.

If there are many items on the job description that are unfamiliar, there are many options to take quick courses in a few days leading up to the interview. This will help you thrive in the technical aspects as well. Sites such as Udemy, Pluralsight, and even YouTube have many tutorials on technical subject matter. Another aspect, if it applies to your interview, is coding challenges. Coding challenges can be very intimidating, due to being watched by the interviewer while trying to write code. A way to practice for this is to use websites such as *LeetCode*. These websites also offer optimal solutions to coding challenges as well. A good idea is to do these often, even when you are not interviewing, in order to become comfortable in these types of case scenarios.

In general, you should have an understanding of what is expected from the role you are applying for, and you should also have an understanding of your strengths and weakness. If coding is not your strength and there is a coding challenge in your future, spend some time studying algorithms and common programming challenges. The interview has a lot of interview and preparation material, and you are just limited by the time you have to absorb it.

> **Pro Tip: Interviewing for Practice**
>
> The best way to improve any skill is by practice and repetition. This is true for interviewing as well. Even if you are not actively looking for a job, you should apply for positions and go through the interviewing process. Each time you do this you will gain additional insight into the interview process; use the feedback you get and apply it to your preparation rituals.
>
> You likely have a shortlist of companies that you would leave your current employer for if you were offered a job. And guess what?! You are not going to be offered a job if you never apply, and likely you haven't applied because you do not feel qualified or have heard stories of how hard the interview process is. Let this be the push for you to go and try. There is a high probability you will fail the first time, especially if it is for a company such as Netflix, Google, Amazon, or Facebook.
>
> As I mentioned previously, practice makes perfect, and in this case, failure will only make you better, and who knows, maybe you will land your dream job at your first attempt.

Beyond doing your research and technical preparation, you should answer some common situational questions, such as what would you do if, or have you ever done this or that. If you search online for interview questions, you can find some common questions and then compile answers for them in writing. Rehearsing these answers will make sure not only that your answer is strong and convincing, but that your delivery is crisp and enticing. You don't want to clear the technical hurdles and then stumble because you had a hard time communicating how you overcame your challenges or how you collaborate across the organization. As with everything else, practice makes perfect.

What to expect

If things go well after you finish the entire interview loop, you may get feedback in as little as 1 week. This could be longer if they have a lot of candidates to interview. If you did not perform well, the loop may end prematurely, and you may get an automated message thanking you but they are moving forward with other candidates. Do not be discouraged. I estimate it takes at least 50 applications to get an offer for a job you may really want, taking 5-10 interviews. Very few people apply for one role and get an interview and an offer. There is a lot of competition for high-paying technical jobs, so be patient and try to stack the deck in your favor as much as you can.

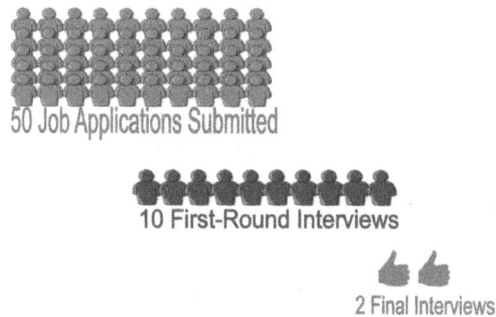

Figure 8.3 – Submitted applications versus final interviews

Normally, after every round, you have the opportunity to ask questions, so feel free to ask about the next stage and gather some early intelligence. If you need more time to prepare, don't rush your interview! Let your recruiter know and reschedule if needed.

If you do get a rejection, do not be discouraged. While it would be great to ask the company that just turned you down regarding the reasons behind the rejection, more often than not, they refuse to go into detail, and it could be a combination of lack of time to address every candidate, or even to avoid some kind of liability. Either way, you can expect an automated rejection email, stating that they are moving forward with other candidates.

Even if you don't get direct feedback, now is the time to analyze the situation for yourself and figure out what went wrong, what when right, and what you can do better.

Sometimes, you can fail to land a job even after you thought you performed great in all the interviews simply because there could be a better candidate. Sometimes a better candidate could be for all sorts of reasons, including a candidate who asks for less pay, a candidate who has been referred, or a candidate with a preferred geographical location. It could be a myriad of reasons, but if you get into the habit of analyzing your performance after every failed interview loop, you can rest assured you will gradually get better. In my experience, you can always know more, but beyond that, you can always communicate better.

The continuous interview cycle

Next, I would like to introduce the concept of the continuous interview cycle, which consists of three phases: **application and preparation**, **interview**, and **feedback**.

Figure 8.4 – Continuous interview cycle

We will start by discussing the application and preparation phase of the interview cycle.

Application and preparation phase

You are applying for roles as well as preparing for upcoming interviews in this phase of the cycle. If you have already received feedback from previous interviews, ensure that you incorporate that feedback into your preparation to ensure your interview technique is continuously evolving and improving.

Interview phase

The interview phase consists of all the interviews that are part of the company's interview process. This phase lasts until you receive feedback, which could be positive or negative.

Feedback phase

This is the stage where you must process the feedback you get from the employer. In the event you are offered a job, you must decide whether to accept the offer, decline the offer, or counter the offer. In the event that you are not offered the job, you may be sent feedback as to why; if you are not, you should do a retrospective of the interview process.

In the next section, we will discuss the tricks of the trade.

Tricks of the trade

The way to get the best results out of your time invested is to make sure you devote the necessary time to preparing for every interview. This can be difficult if you have a lot of interviews going on and you are still working, but consider that you can be disqualified over a number of things, but you won't know until after the effort has been expended. It can be tough to learn that you didn't make the team over a small issue after spending several weeks interviewing. Don't interview for roles you are not really interested in, or that you don't think are worth the extra preparation. Do yourself a solid and stack the deck in your favor.

In the next sections, we will go through things that I have found to be effective in the interview process as well as things that I have found to be ineffective and detrimental to your success. First, take a look at the visual representation seen in the following diagram:

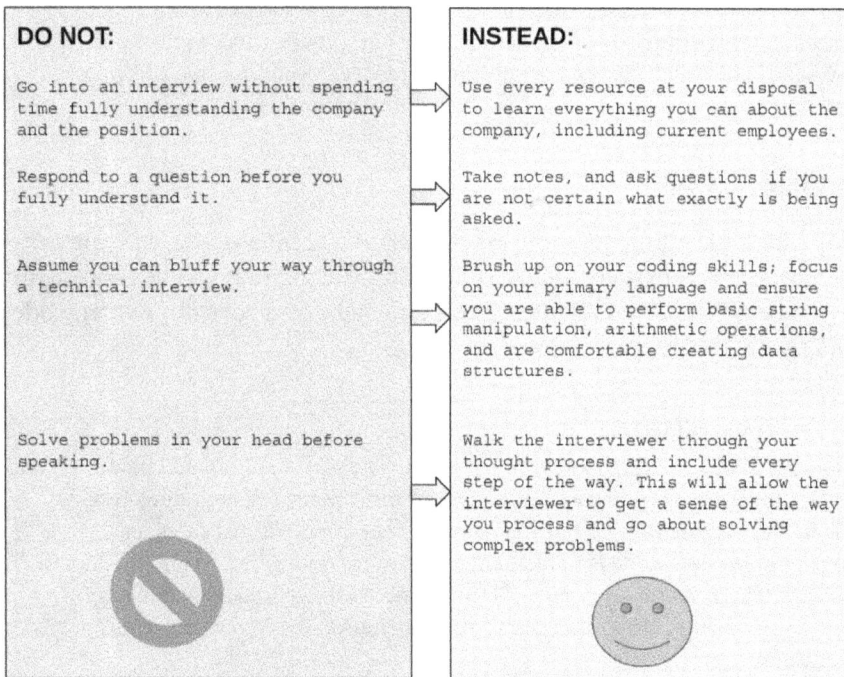

```
DO NOT:

Go into an interview without spending
time fully understanding the company
and the position.

Respond to a question before you
fully understand it.

Assume you can bluff your way through
a technical interview.

Solve problems in your head before
speaking.
```

```
INSTEAD:

Use every resource at your disposal
to learn everything you can about the
company, including current employees.

Take notes, and ask questions if you
are not certain what exactly is being
asked.

Brush up on your coding skills; focus
on your primary language and ensure
you are able to perform basic string
manipulation, arithmetic operations,
and are comfortable creating data
structures.

Walk the interviewer through your
thought process and include every
step of the way. This will allow the
interviewer to get a sense of the way
you process and go about solving
complex problems.
```

Figure 8.5 – Do not do that; do this instead

In the next portion of the chapter, we will cover mistakes that candidates often make during the interview process.

Common mistakes

The following is a list of things that have been used by me and proven effective during the interview process. It is not a complete list and I encourage readers to go and discover insights from others, so as to give yourself the best opportunity for success with your interview:

- Not fully understanding the question:

  - This happens very often in all types of interviews, not just technical interviews. Often, the interviewee will hear a few words and start coming up with solutions in their head and miss the rest of the question and end up getting it completely wrong. Always wait to hear and fully understand the question before going to the solution phase.

    > **Pro Tip: Take Notes**
    >
    > The best way to ensure you do not miss anything is by taking notes. You can use a notepad and pen if the interview is in person or a notepad app on your computer if it is virtual. This will demonstrate your interest and, at the same time, allow you to fully grasp the question before responding, while allowing you to ask any question you may have before answering as well.

- Not writing edge cases:

  - Many times, in coding questions or challenges, the interviewee will not write any edge/corner cases in their code. Remember, your solution will be running through pre-generated test cases, and it will certainly fail if you have not written code to manage edge/corner cases.

    > **Pro Tip: Practice TDD**
    >
    > In the first chapter of the book, *Chapter 1, Career Paths*, we discussed how DevOps incorporates many extreme programming practices, including **Test-Driven Development** (**TDD**), which has the program write test cases that will fail until proper code is implemented. If time allows, which would be the case with take-home assignments, TDD practices can be used to ensure edge cases are not missed while also demonstrating your knowledge of extreme programming practices.

- Unfamiliarity with the language:

  - Interviewees will want to write code in Python during interviews since it is like pseudo code. This is perfectly fine if you choose to do so, but please be sure to be sufficiently familiar with the language of your choice so that you know how to perform basic string manipulation, arithmetic operations, and are comfortable creating data structures. If you are not very familiar with the language of your choice, this will leave the interviewer thinking that you are not familiar with writing code or that you do not write code very often.

    > **Pro Tip: Sign Up for Daily Coding Problems**
    >
    > Sign up to receive a coding challenge problem in your email each day at `https://www.dailycodingproblem.com/`. The following day, you will be sent the solution. I have used this, and it has helped me with coding interviews, as well as enabling me to feel more comfortable with pair and mob programming.

- Being quiet:

  - Often, when interviewees are thinking, they will do it quietly. This is the opposite of what is expected. Usually, the interviewer just wants to see the way you think and the method by which you solve problems. Being quiet leads to a misrepresentation of assumptions by the interviewer. Instead, work through the problem verbally and ensure the interviewer is included in your thought process.

    > **Pro Tip: Don't Pretend to Know Material That You Do Not!**
    >
    > Often, those that are getting interviewed panic when they are asked a question and they do not know the answer or are completely unfamiliar with the subject matter. In the technical world, new technologies are being implemented daily, so you will not know everything. This is NORMAL. It is almost always better to just state "I am not familiar with this, but it sounds really interesting to learn about," rather than try to make something up. Remember, the interviewer is probably an expert with that technology, and they will know immediately if you are bluffing.

Things that do work

The most important thing when it comes to negotiation is to do your research:

- On the company, on the role, on the skills required, on the hiring manager, on the news. Even research on the interview process itself can give you the edge you need to prepare. Research, research, and undertake more research to get the edge that other candidates do not have.

> **Pro Tip: Find an Inside Source**
>
> The best way to get information on a company is by chatting with someone from that company. LinkedIn makes it easy to find connections who work at specific companies who are within your network.

- Once you are in the thick of your job search, you can expect to have interviews with more than one company simultaneously:

  - Keep a spreadsheet tracking who the recruiter and hiring manager are at a minimum, as well as any facts you may need at a moment's notice. It's also useful to keep track of where you are in the interview process, as well as what level of remuneration was discussed in an earlier conversation.

> **Pro Tip: Be Consistent**
>
> You sound disorganized and unprofessional if you give two interviewers two different answers to a question. The most common place this occurs is with remuneration. The recruiter will almost certainly ask you at your initial interview what you are expecting in terms of remuneration and it is important you remember this number and don't inflate it later in the process.

- Applying the principles discussed in the continuous interview cycle section:

  - Always stay fresh and comfortable with interviewing by interviewing regularly, taking the feedback you receive from the interview, and applying that as you prepare for your next interview.

- Be comfortable with what you do not know:

  - It is completely acceptable and expected that you will not know everything. Be upfront with interviewers as regards your strengths and weaknesses.

> Pro Tip: Be Comfortable with What You Don't Know, but Be Open and Interested in Learning New Things.
>
> At this point in the book, it is obvious the value I place on individuals who prioritize continuous learning and self-improvement. Believe me, I am not the only person in a position to hire individuals who have the same outlook. In some cases, someone with a fresh outlook without prior knowledge is what a team needs.

Summary

In this chapter, you gained a lot of insight into the interview process, as well as how to prepare for the distinct phases commonly encountered within.

We also talked about the various stages and what you can expect, separating the details for an individual contributor as well as for a leadership role.

We talked about what to expect and how to get the most out of every interview loop, even when you do not end up with an offer in hand.

Finally, we covered some tips and tricks on what works, and what does not work, in an interview setting.

We also covered diverse types of interviews in some detail, but we will cover that in more detail in the coming chapters. In the next chapter, we will cover both typical and non-typical interviews and how to navigate and be successful in both types.

9
Interviews Step by Step

In the previous chapter, we talked about the interview process without going into too much detail. In this chapter, we will continue this discussion by diving deeper into the topic, including the different steps, the different people you may speak to, and how to best prepare for each level. To do this, I will highlight specific examples and extra details regarding typical and non-typical interviews. From coding challenges, design, situational, and IQ tests to everything in between, I will try to cover everything and give you tips to become bulletproof, regardless of the type of interview you may encounter.

In this chapter, we will cover the following topics:

- Typical interview walk-through
- Non-typical interview walk-through

Typical interview walk-through

The following diagram breaks down the interview process into three phases – the initial round, the technical assessment, and the follow-ups:

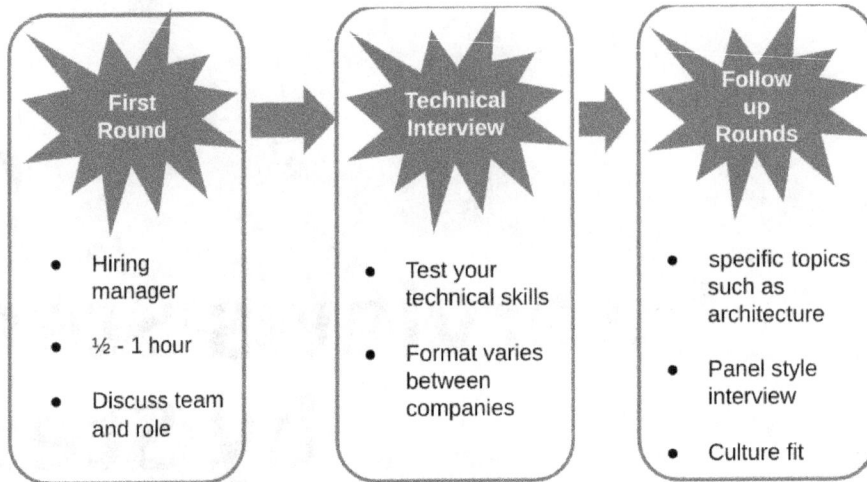

First Round
- Hiring manager
- ½ - 1 hour
- Discuss team and role

Technical Interview
- Test your technical skills
- Format varies between companies

Follow up Rounds
- specific topics such as architecture
- Panel style interview
- Culture fit

Figure 9.1 – Interview stages reviewed

We will cover these in more detail in the following sections. Simply put, each round is equally important to the outcome of the interview, so bring your A-game to each round.

First-round interview

The first round is when you speak with the hiring manager or someone doing a similar function. You will have calls with recruiters or HR beforehand, but for the purposes of this walkthrough, these are just prerequisites to the first round interview call.

What is Considered a First-Round Interview?

The first-round interview is considered the interview you have with the hiring manager, or the individual who will assess your qualifications and fit for the role. It is the interview that occurs after you have completed the human resources screening call successfully.

For the first round, you must focus on doing homework on the company, check the LinkedIn profile of the interviewer (they will be your direct supervisor) for clues, and make sure you read the job description a few times.

Besides background information, you should also focus on conveying your soft skills, primarily going through your experience cohesively and succinctly, and thinking how your previous experience may benefit your potential employer.

In this round, they will typically go into details of the current challenges, so feel free to ask ad hoc questions throughout and save some for the end, where they will expect some questions from you. You can use this question section to tie in some of your previous experiences and explain how you have solved similar challenges in the past. This helps leave them with the impression that you know how to solve the challenges that they are facing and it is a wonderful way to end the call.

If you are interviewing for a leadership role (and even if you are not), you should also ask about the team that you will be working with.

Do not mention compensation, benefits, or anything like that unless the interviewer explicitly brings it up.

The following diagram specifies some dos and don'ts for first-round interviews:

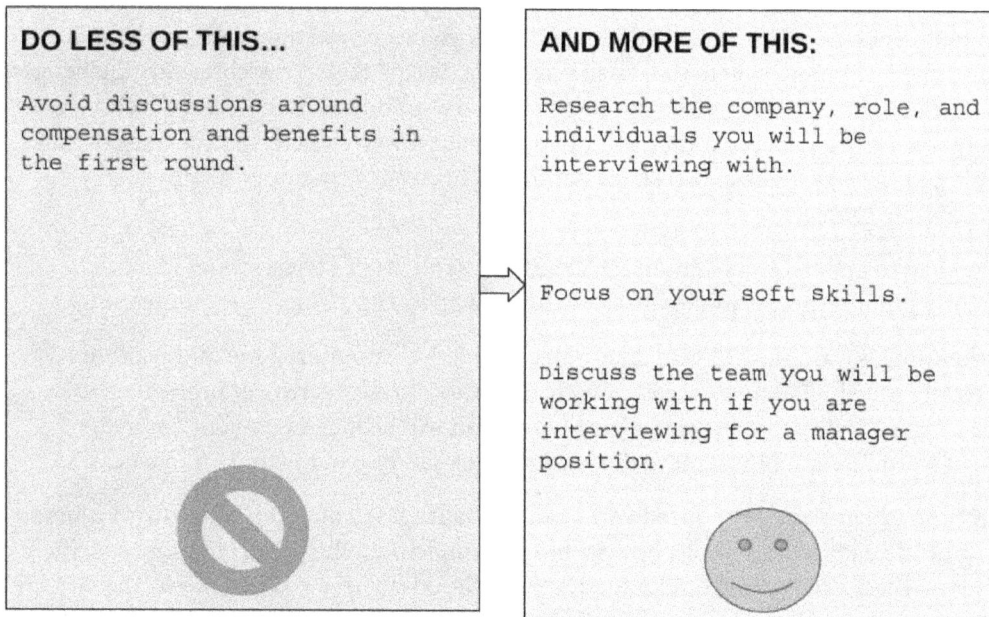

Figure 9.2 – First-round interview dos and don'ts

Make sure you ask plenty of questions! This is the best time to show that you did your research, that you are ready to get started, and that you are passionate about the job. For a lot of companies, how interested you seem in the role is important. Your technical qualifications might be there, but if you look or sound bored through your interview, you may miss your shot!

Technical interview

This is the most critical part of the interview as the goal is to determine the depth and breadth of your technical knowledge; there is only so much preparation you can do between the interview with the hiring manager and this phase. It is best if you do not misrepresent your skill set so that you are not asked about things you have no experience or proficiency in. If you do encounter something you do not know, explain that you have not worked in that area in the past, but that you are interested in doing so. If you pull something out of left field, you may cause irreparable harm to the trust relationship you are trying to build.

Different companies will employ diverse ways to gauge your technical skills, from most common scenarios, where one or more people ask you technical questions, to the more sophisticated instances, where they may give you a design, code, or architecture challenge to do ahead of time, only to discuss it in person once submitted. Simple technical screens usually consist of a panel or multiple individual interviewers, where you can expect your technical knowledge to be explored for depth and breadth. Remember, anything in your resume is fair game.

Once, I had to architect a system in the cloud and send proof of everything I had done, effectively granting access to the infrastructure I was hosting. I was later quizzed on it.

Another time, I was asked to do a take-home problem that required me to use a tool I specifically stated I had no experience with. The purpose of this type of problem is to highlight your ability to learn something new – you will be judged on your ability to creatively produce a solution more than your understanding of the tool, in most cases.

Many times, I have also been faced with coding challenges, which can range from general coding proficiency (for example, how would you implement this *XYZ?*) to more specific problems. These can range from honor system solutions, where you just have a day or two to return the test and there is no way to know if you googled the answer, to sophisticated third-party proctored tests, where a real person will watch you code. Some companies may want to watch you code live and ask you questions as you work on your solution.

The more demanding coding tests are typically for software engineers, but it is common for DevOps engineers to be expected to have some programming proficiency. The bottom line is that it pays to know at least one language well enough to implement common algorithms and data structures.

> **Pro Tip**
> DevOps engineers are required to be excellent software engineers and programmers; some of the most proficient coders I have known have been fellow DevOps people.

Other types of technical tests might be more domain-specific, such as writing Terraform modules or shell scripts to automate a task. Whatever the test is, make sure you gather as much intelligence on it before you start and expect to have your solution explored at length in a follow-up conversation.

A design interview or challenge is a little bit different. It focuses on creating a new application or service and building the respective infrastructure and services. It may require you to make decisions around databases, security, or even areas more traditionally focused on application development. Some companies will test to see how well you understand the entire application stack. It is important to have some high-level understanding, especially in the cloud. You should know how to segregate and secure networks, as well as how to prevent the internet from accessing your backend and databases. You should know how to do authentication and create security groups and understand networking well. Distinct roles will have different focuses. Some will want to focus more on containerization and may go deeper into Kubernetes. Some may cover serverless. Getting a few certifications under your belt ensures you have a broad understanding of services that you may not have used professionally. They also help differentiate you from the competing interviewers.

Even in technical rounds, it is important to end the interview by asking questions. I like to ask technical folks how well they enjoy their job, or what challenges they see. It might be a different answer from the first-round call.

Additional rounds

You should work in tandem with your recruiter to know exactly what to expect from each round. When in doubt, the most foolproof method is to look up the individual on LinkedIn and see what they do. In my experience, if they are in a different area than you, such as **quality assurance** (**QA**) and testing, it is usually a cultural or team fit type of interview, which may have a few situational questions. If it is in your area, they may want to know more about your specific experiences and see if you match the challenges they face.

Even if you aced the technical interview, do not relax! You can still fall short by not showing sufficient enthusiasm, or by not finishing your last interviews strongly. Remember that even if you did very well, someone else could have done better, so always give one hundred percent until the final round is complete.

If possible, wait some time and ask for feedback. If you do not get the job, ask for feedback too. Sometimes, they will not give you anything more than a generic rejection, but sometimes, you get the good stuff, which points out a key flaw in your interview process, such as being *too hands-on* or *not hands-on enough*, or if your enthusiasm did not match their expectation. I had one person tell me I was not showing enough emotion and it was off-putting for them.

Besides formal feedback, try to analyze your performance and give yourself some feedback. Were your answers as smooth or succinct as they could be? Did you have any knowledge gaps?

The following diagram shows different types of technical interviews:

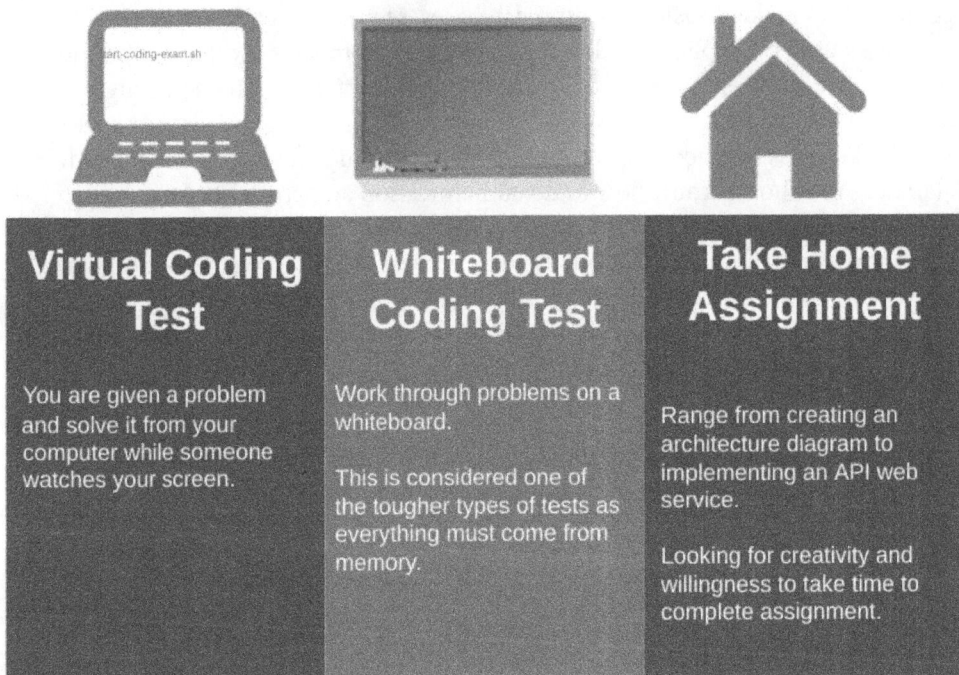

Virtual Coding Test	Whiteboard Coding Test	Take Home Assignment
You are given a problem and solve it from your computer while someone watches your screen.	Work through problems on a whiteboard. This is considered one of the tougher types of tests as everything must come from memory.	Range from creating an architecture diagram to implementing an API web service. Looking for creativity and willingness to take time to complete assignment.

Figure 9.3 – Technical interview types

In the next section, we will discuss the offer stage of the interview process.

Offer stage

Congratulations – you cleared the loop and overcame all the hurdles, and their recruiter or HR person has reached out to let you know they would like to make you an offer! Typically, they make a verbal offer and if you accept verbally, they will extend a written one, which usually has an expiration date – around 48 hours to a week.

Now, if you have been interviewing with multiple companies concurrently, you may be in the process of receiving more than one offer and potentially be able to negotiate further. Be careful with overplaying your hand, since companies may rescind their offer if they think you are just using them to get a better offer elsewhere or to improve your existing situation at work. The point is that you should tread carefully as the job is not yours yet, so treat this as the final stage of the interview.

You should focus on being very enthusiastic about the role, even while you negotiate. Also, try to be flexible when it comes to improvements in your offer, since they may have more flexibility in one area than another. Be wary of telling all your other opportunities that you have received an offer too early, as it may prove counterproductive. Do not mention companies by name or give away too much in terms of the specifics. If you get two or more offers, then negotiate in earnest, but only do so if you are willing to walk away from the opportunity you are trying to improve.

Sometimes, people will take the offer back to the company they work for and get a raise. This might be a good short-term way to get a raise out of band but it can also generate negative feedback and, depending on the employment market, they may look for some insurance in the future so that it does not happen again. Take this case by case. If you have not received a raise in several years and you get an offer, negotiate on both ends. Just try to manage your options so that you are still on good terms with both companies after deciding.

Non-typical interview walk-through

While most interviews tend to fall into a pattern, occasionally, you will encounter something out of the norm that is worth highlighting. In this section, we will cover pre-screening tests, out-of-the-box designs, and example questions such as the infamous *Tell me a time when* formatted question.

Tests

First, let's start with pre-screening tests. Some companies want to know your personality type and will want you to take a personality test ahead of time. This can range from one page where you evaluate five words you identify with, to long 15+ question exams.

Sometimes, this is paired with a cognitive test, which is like a traditional IQ test, albeit with a lot less time dedicated to this. I have taken the **Criteria Cognitive Aptitude Test (CCAT)** several times, and it is 50 questions in 15 minutes. Only once did I finish all 50, and there was a generous amount of guessing to get that done. The questions range from math and logic questions to spatial and verbal questions such as analogies or antonyms. It requires all your focus, so try not to take these types of tests when you are tired or may be distracted.

I have even heard that some private equity companies have everyone in the companies they invest in take these tests! There is not much you can do to prepare for these types of tests, but you can stack the deck in your favor by making sure you have 15 minutes of uninterrupted focus, at the time when your brain operates at peak capacity. Since you are not penalized for guessing, you should always dedicate the last minute to guessing all the remaining questions.

If you do not have experience with these types of tests, try finding a practice test:

Sample CCAT Verbal Question:

Choose the word that is most nearly OPPOSITE to the word in capital letters. LENGTHEN

A. abdicate

B. truncate

C. elongate

D. stifle

E. resist

Show correct answer

Figure 9.4 – Sample CCAT question – 1

The preceding screenshot shows an example of a verbal question from https://www.criteriacorp.com/. The following screenshot shows an example of a math question, also from https://www.criteriacorp.com/:

SAMPLE CCAT MATH QUESTION

A group of 3 numbers has an average of 17. The first two numbers are 12 and 19. What is the third number?

A. 17

B. 19

C. 20

D. 23

E. 30

Show correct answer

Figure 9.5 – Sample CCAT question – 2

Besides personality and cognitive tests, you have programming and design tests, which we have talked about previously. For traditional programming tests, `https://leetcode.com/` is a suitable place to start, as is `https://www.hackerrank.com/`. For books, check out *Cracking the coding interview*, by *Gayle Laakmann McDowell,* which contains a lot of tips specific to companies.

Out-of-the-box design

For design, this is a little bit more complicated, but there are a few good books that go in-depth into system design and can help you not only on offline tests but also in working through a design problem live with an interviewer. For cloud-related design, check out reference architectures from the cloud provider itself. If you study a few samples, you will get a handle on what is common to a lot of scenarios and be able to think of solutions more rapidly. I recommend that you become acquainted with at least one tool for diagramming and try to design a sample three-tier application in the tool before you are forced to do this live.

You could encounter a non-conventional problem, such as the famous *how many manholes in New York* or something like that. I was asked what I would do if I was shrunk down and put in a blender. I was also asked to design the process for a map application from the ground up, in an underdeveloped country lacking infrastructure. A company may also ask you to design one of their services, improve it, or create something new. A lot of big tech companies may screen you this way, regardless of your specific technical role, so saying you are in *DevOps* may not preclude you from this.

The following figure contains some non-conventional questions:

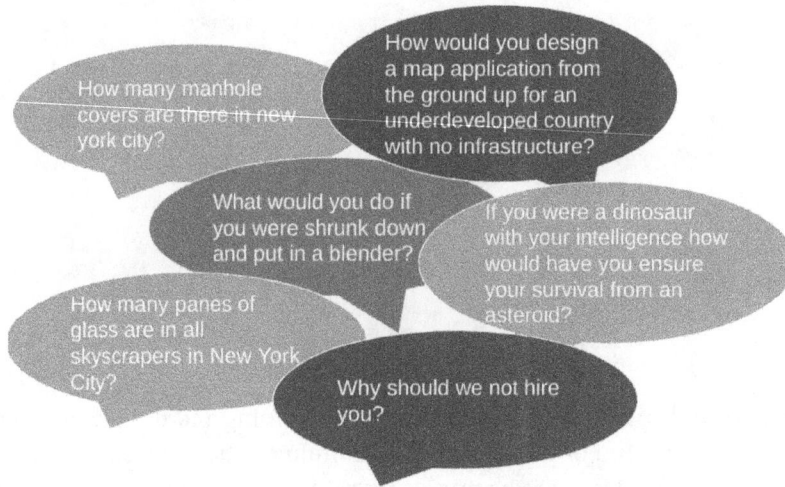

Figure 9.6 – Ridiculously hard interview questions

Thankfully, the most insane of the brain teaser types of questions have become extinct for the most part and are no longer seriously being used to screen candidates. As always, ask your recruiter what to expect and if the company is large enough, the internet is bound to have additional intelligence on what types of questions or design problems they can throw at you. What matters the most in these types of exercises is that you can convey your thinking, that it is clear and logical, and that you consider exceptions and edge cases when devising solutions.

Tell me about a time

One of my favorite cloud providers is famous for asking situational questions that start with the *tell me about a time* mantra. These types of questions will ask you about your professional experience, how you overcame adversity, and how you deal with every possible professional situation. The catch is that you should do it in the STAR format – **situation**, **task**, **action**, and **result** – as shown in the following diagram:

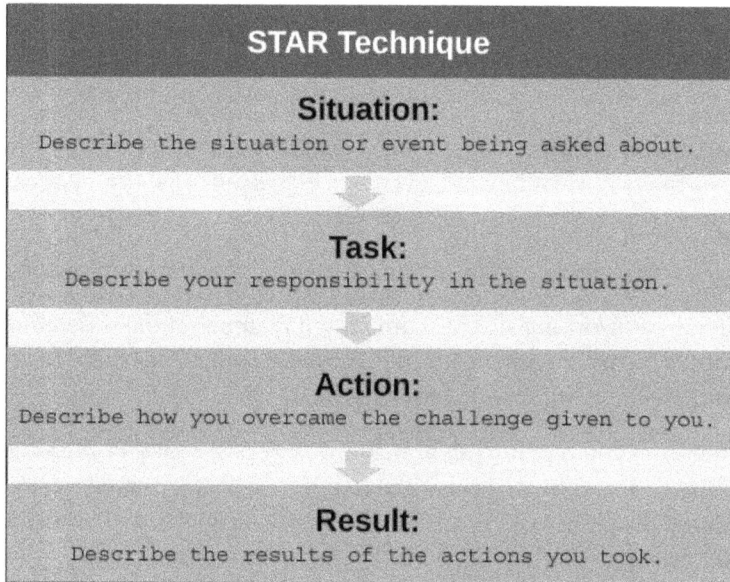

STAR Technique

Situation:
Describe the situation or event being asked about.

Task:
Describe your responsibility in the situation.

Action:
Describe how you overcame the challenge given to you.

Result:
Describe the results of the actions you took.

Figure 9.7 – STAR technique

What happened, what did you do, and what was the result? Here, you should put less emphasis on the company or team and focus on your contributions. The ideal candidate is a maverick that will go against the current norm and challenge the status quo to get the best outcome for yourself, your team, or your customers.

One of the challenges here is that your own experiences may not be quite as interesting or as heroic, or that you may not have an arsenal ready to use at a moment's notice. This might be a pain, but the solution is simple. Create a document and start working on stories, answering sample questions in the STAR format. You do not want to repeat stories in the same interview loop with the same company, so make sure you can pivot to a different story if there is a similar question that you have already used your story for.

Tell me about a time when you had to work with someone difficult to work with?

ST: I had to work with a teammate who was difficult to work with.

A: I spent extra time getting to know this person and our relationship improved.

R: This person is now great to work with.

Now, this story does not have any flavor, but I stripped that out to show you the core mechanic.

So, how can you make your story better? You can add data and numbers. Adding a quantitative feel to your story makes it more memorable and impressive (especially compared to the stories of other candidates that do not have concrete data to share).

Typically, the R portion should be positive, but it does not always have to be completely positive. If it shows that you went beyond, even a negative outcome can be seen as a positive answer, as shown in the following example:

Tell me about a time when you had to work with someone difficult to work with?

ST: As a junior developer working at ABC company, a member of my squad did everything they could to cause drama with me. They went out of their way to try and drive distance between myself and other squad members.

A: I was new to the team and the individual who I was having a tough time with had been a member for 4.5 years. I invited the individual to lunch so that we could get to know each other better. Additionally, I made it clear I was in no way trying to take over their team and only assist as needed.

R: 2.5 years later and drama dude and I are peers, and we tend to have a strong working relationship. We respect each other but also are extremely competitive with each other as we are constantly trying to push each other just a little bit further. Honestly, without the relationship we have, I would not have been promoted to a senior engineer on the same team as him. He is one of the individuals who recommended I apply for this position.

Be aware that this type of question can apply to both technical and non-technical scenarios. There is a stronger bias toward leadership questions the more senior you are, but you should be prepared to frame any answer in the STAR format.

Mistakes to Avoid

When answering questions, avoid using *we* or *us*. Instead, focus on what you did. Even if it was a team effort, focus on what you provided for the solution.

Another mistake I see candidates make is answering a question with a simple yes/no instead of using the opportunity to sneak in some more relevant information about themselves. Depending on the position you are interviewing for, you may be allowed to redeem yourself if the interviewer asks you directly for a more specific example.

To recap, find a list of situational interview questions and start answering them, keeping your questions and answers handy in some cloud document. Then, refine your answers so that they are snappy and easier to share.

Summary

In this chapter, we covered traditional and non-typical interview walk-throughs in more detail.

We covered the traditional stages of the interview process and focused on the technical assessment stage, discussing diverse types of tests and screening, as well as how to prepare.

We also discussed non-traditional interview scenarios, such as cognitive and personality tests, situational exercises, and design questions, including out-of-the-box brain teasers and others. We stressed practicing tests and writing out our answers to the most common situational questions ahead of time and keeping them handy. Practice makes perfect, and interviewing is not an exception!

In the next chapter, we will discuss tips and tricks for applications and the interview process.

Section 4: Tips, Tricks, and Interviews

The authors' combined experience of 25 years working in the field of DevOps has led to the final section of our book. It provides exactly what the title says – tips, tricks, and interviews useful for DevOps professionals.

This section comprises the following chapters:

- *Chapter 10, DevOps Career: Tips and Tricks*
- *Chapter 11, Interviews with DevOps Practitioners*

10
DevOps Career: Tips and Tricks

In the previous three chapters, we defined DevOps, covered the various paths you can take, and laid a foundational plan you can follow during the application process, as well as each stage of the interview process. In this chapter, we will focus on the application and interview process.

In this chapter, we will cover the following main topics:

- Tips for transitioning to a career in DevOps
- Tips and tricks – things to avoid during the interview process
- Tips and tricks – things to do during the interview process

Tips for transitioning to a career in DevOps

This section revolves around my DevOps journey. I will start laying the groundwork by telling the story, followed by tips that can be extracted from it.

Personal DevOps journey

In 2005, the decision to attend college for mechanical engineering at the *University of Minnesota* was made. In college, I needed to take a C++ computer science class. This class was the extent of a software-related course that could be taken in college; however, the course got me inspired to continue exploring programming and what could be done with it. This fascination led to me purchasing a microcontroller board, similar to what is known as an Arduino. I ended up creating a hosted web page on a web server, which lead to a dorm room that could be monitored from a computer.

I took a lab that focused on automation and robotics in my senior year in college, which inspired my interest in automation. The professor for my course also spurred my interest in automation. In the lab, we decided on a product we had to manufacture with no human interaction – we chose a wooden cribbage board. We were broken into groups of three and assigned a Fanuc six-axis robot, as well as a stage of the manufacturing process we were responsible for. My team was responsible for the packaging cell. After the cribbage board was created, it needed to be placed in a box, labeled, and shipped. I became very interested in the entire process and ended up spending far more time in the lab working on my cell, as well as helping other cells integrate with ours. The professor became the first academic mentor I had and was the reason I ended up taking the job I did out of college.

My first job out of college was working as a manufacturing engineer, a job I did not enjoy but took because of the exposure I was able to get with automation in an industrial setting. I sucked in as much as I could from my senior colleagues who worked in the automation department; one individual told me that it was impossible to master something quickly when you spend only a small portion of your time working on it. He recommended that I buy a used microcontroller to play around with on my own. I did and ended up setting up a home lab that grew into home automation tasks, and eventually into a fully automated home.

My time as a manufacturing engineer ended and I started my career as an automation engineer working in the oil and gas industry in western North Dakota. During my tenure as an automation engineer, I was exposed to **supervisory control and data acquisition (SCADA)** systems. The engineers working on SCADA systems were developing software in Java and using Maven, Subversion, and Eclipse. I began researching software development in my free time and began learning new programming languages and tools in my spare time. Once I felt comfortable with programming, I began applying for jobs as a software engineer for the same company. Eventually, I landed my first software engineering role.

Software engineering was crazy fun – there was always something new to learn and a challenge to solve. I had a knack for picking up programming languages pretty quickly and wished to work with a more diverse code base. Another thing that I was struggling with was balancing work and life, or a lack thereof. I lived in Minnesota but needed to commute to western North Dakota frequently. I knew several individuals who worked at United Healthcare that continually encouraged me to apply. I applied to half a dozen software engineering roles I thought I was a good fit for.

I also applied for a role as a technical agile coach, a role whose description sounded fascinating, but I did not feel I was overly qualified for. I had no experience using Agile, being a coach, or experience with DevOps. I ended up landing the role! Being a technical coach involves working with teams to help them further their DevOps practices. Overwhelmed did not begin to describe how I felt on the days leading up to my first day. I spent every hour I was awake studying DevOps and technical practices, only to get more discouraged as I learned how deep a rabbit hole the field was. On my first day, my manager sat me down and told me this:

"I know you don't have experience with DevOps or Agile; there were more qualified candidates than you, but you were chosen because of your desire and hunger to learn."

This individual eventually became what I considered my first mentor. Under this individual, I grew both my technical skills as well as my soft skills. This individual helped me become more confident and encouraged me to attend Toastmasters to help with my public speaking. When I did leave their team, it was to tackle a new challenge – that is, building and leading a DevOps team of my own.

The previous story was a goldmine of information; some of you may have missed some things, so let's break it down piece by piece.

Stay on track, but entertain your interests

While in college, my degree did not allow me to take additional software courses, so I decided to follow up on this in my own time in a way that would not affect the time it took me to get my degree. This is not always an option; sometimes, you need to go all in. Unless I am certain of something, I usually hedge my decisions – that is, I continue doing what is working in parallel to my new idea/process. A notable example of this is as follows:

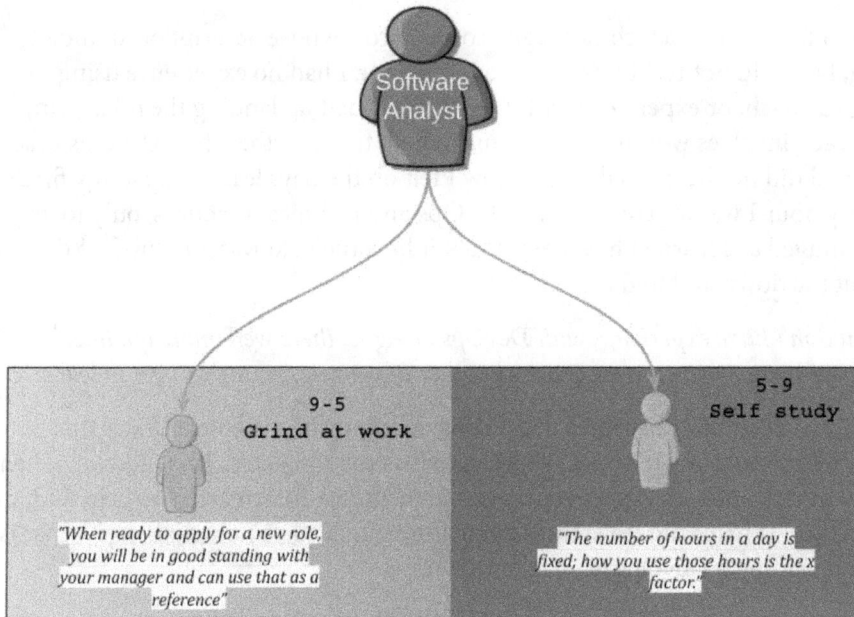

Figure 10.1 – Use your day appropriately

In the preceding example, the individual in question has a job as a software analyst, but they would like to move into a higher-paying software engineering role. They have heard of software boot camps and are considering leaving their job to attend a 6-month boot camp. I strongly caution against doing this; first, you are not guaranteed a job after you graduate. You could go through the boot camp, only to come out and not be able to land a job. The second reason I do not feel it is a great idea to leave a job without a solid job offer in hand is that you will go into debt for tens of thousands of dollars. Instead, I recommend staying at your current job and taking advantage of the tuition reimbursement your company offers. Use these funds to take programming courses, get certifications, and even pursue an online degree. On top of this, something you can do even if your company does not offer tuition reimbursement is research and study on your own. Think of the amount of time you spend watching Netflix and playing video games; use a portion of this time to learn new things that will help you land your next job. I respect candidates who are self-taught and speak with many other leaders who have similar opinions.

Life is busy; prioritize and focus on things important to you

In college, I was busy – too busy, honestly. I was volunteering as a ski patrol, taking 16 credits, working part-time, and trying to have a social life. It was not until the robotics lab that I realized I was going to need to reprioritize if I wanted to deliver a demo that I was proud of. In other labs, I was fine delivering something that was not my best work, so long as I would get good marks. I was extremely interested in the robotics lab and decided I wanted to get the most out of it so that I could use it in my career down the road. I prioritized the lab at the expense of having to give up being part of the ski patrol. Another example of this can be seen in the following diagram:

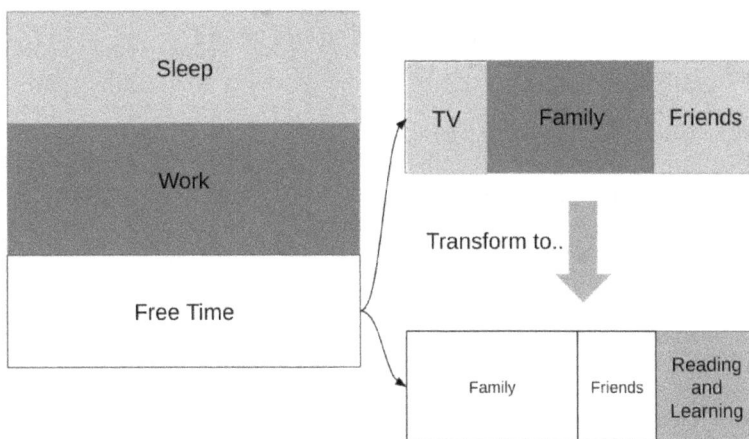

Figure 10.2 – Reprioritizing your time

As you can see, there are a fixed number of hours in a day; you should also try to get adequate sleep, and you need to put in your time at work so that you can continue to perform at a good level. This leaves you with your free time; you are in control of how you spend this bucket of time. I recommend sitting down and determining areas that could be cut out or cut back to free up your time for activities you wish to prioritize.

Opportunity is often disguised in a deceptive facade

In the example at the beginning of this section, I discussed how I took a job as a manufacturing engineer because I knew I would have the opportunity to work with the automation and control systems department; what was not said was that it took me a lot of time getting to know the company as an intern to discover this opportunity. This has happened several times in my career; in some instances, I have done upfront research and made a good decision; other times, I missed a fantastic opportunity. Take, for example, the following diagram:

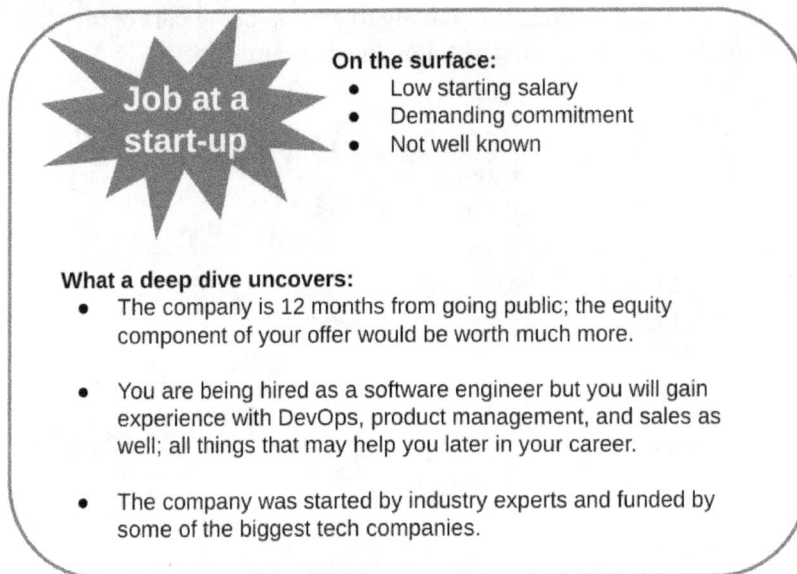

Job at a start-up

On the surface:
- Low starting salary
- Demanding commitment
- Not well known

What a deep dive uncovers:
- The company is 12 months from going public; the equity component of your offer would be worth much more.

- You are being hired as a software engineer but you will gain experience with DevOps, product management, and sales as well; all things that may help you later in your career.

- The company was started by industry experts and funded by some of the biggest tech companies.

Figure 10.3 – Looking beyond what is seen on the surface

The preceding example is an extreme case for example's sake, but the logic of doing your research ahead of deciding on a major event still holds true.

Making an internal career change is sometimes easier than changing jobs and companies at the same time

When you are working for a company, there is a good chance you are building up a strong rapport with your management. If you have decided that you wish to pursue another career track, it is always advisable that you speak with your current employer first. In this example, another career track is referencing a shift away from the type of work you are currently doing. The example we will use to demonstrate this is of a software engineer moving to a career as a DevOps engineer:

Internal Path

- You are able to discuss your career aspirations with your leaders, who already know your talents.

- You leaders can better prepare you for a new role by giving you work more related to a role in DevOps.

- You are able to utilize others within the organization as references.

"As a software engineer my goal is to transition into a role focused on DevOps"

External Path

- You may be overlooked because of lack of experience as a DevOps engineer - you will be required to demonstrate your skills with a technical assessment.

- You need to prepare for the role you are applying to by researching as self study.

- If there are internal paths for a career move and you are in good standing with your current leadership, ask them if they would be a reference for you.

Figure 10.4 – Internal versus external job change comparison

There is one thing I am not trying to say here – that applying for external jobs is not a good option. Applying to external jobs has worked for me many times in the past. Sometimes, applying internally shows quicker and more favorable results, as shown in the preceding diagram.

Transitioning to a new career path requires new skills – skills you may have but that will not be evident unless an individual knows you personally. This is one reason transitioning internally can be easier. Another reason is that internal career moves can be made in stages or as part of a plan if you discuss them with your leadership. Finally, internal references for a job are gold. If you can apply for a job and have several people who can recommend you for the job, your chances of getting the job increase exponentially.

Apply for roles you find interesting, even if you do not meet all the requirements

Apply for jobs you want, not ones you are completely qualified for. This will prove ineffective most of the time, but I have landed a job this way, as have many of my connections. Companies that use automated filtering mechanisms will filter you out if you do not meet all the requirements; one way to get around this is by including the time you have spent using said software on personal projects, not just professionally. As a hiring manager, I am not going to overlook you because you are missing 2 years of experience with a certain language or tool.

The problem with setting years of experience requirements for jobs is that it only works if everyone learns at the same pace. However, most hiring managers are aware that this is not true.

Things to avoid during the interview process

In the previous section, I told a story about my successes and how they helped me get into DevOps. Now, I would like to spend a few pages reminiscing on my failures, or more so my failed attempts to land a job. To start, we will travel back to 2008.

Avoid providing inaccurate or misleading information when applying for a position

The time incorrect information cost me a job.

In my junior year of college, I began looking for internships. I applied to one job that required applicants to be pursuing a degree in computer science; I was pursuing a degree in industrial engineering. When the question came up on the questionnaire asking if I was currently enrolled in a computer science program, I answered yes. A week later, I received a call from the recruiter and we began discussing the job; it was a very short discussion that ended with me being sent a link to a technical aptitude test. I must have done OK on it because I made it to the next round, which was onsite. The company paid for me to fly out, as well as put me up in a hotel. I was required to bring my transcripts onsite, which I went through in my first meeting. I was then asked, *I thought you were a computer science major?*. Needless to say, the encounter ended with me having a broken ego, which could have been avoided if I would have followed the golden rule of applying for jobs: always provide accurate information. If I had been upfront, I may still have been offered the opportunity, but because I was not honest in my initial application, I was put on the hard pass list.

A rule that should always be followed is being honest when applying for jobs, as well as on your social profiles. In my case, I knowingly provided false information, but unintentionally providing incorrect information can have the same results. Looking back, there is one word I would use to describe what I did: selfish. I chose to put my wants above what was required by the company. In doing so, the company wasted money interviewing me, flying me out to meet in person, and providing me with a hotel. It was a huge betrayal of trust when they learned I knowingly provided false information. I would not have hired me either – I may have had a chance had I been honest. Avoid going to an interview without fully understanding the requirements of the position.

Did you even read the job description?

A candidate passed the recruiter screening for a DevOps role on my team and quickly moved on to the face-to-face technical interview. For every question I asked the individual, they responded with examples where they used Python to solve a similar problem. At the end of the interview, I gave the candidate this feedback:

Though you seem highly skilled in using Python, we are looking for an individual who has more hands experience using Golang; would you be willing to go through your experience working with Golang?

The candidate answered truthfully that they did not have experience using Golang outside of making slight code modifications. Unfortunately, the candidate did not get the job and could have saved their time if they had fully read the job description: *Senior DevOps Engineer – Golang Experience.*

In the preceding example, the candidate could have avoided an awkward conversation if they had spent a little more time researching the role and requirements instead of jumping into preparation right away. It was clear that the individual was a strong engineer, but on the given team, the ability to mentor others in Golang was a requirement that was discussed in the job description. It is always OK to ask the recruiter if the job description is vague or if you are uncertain about something.

There are four areas where you must come into an interview prepared:

- **Culture**: Understanding the culture of a company will help you get through the initial screening process by a recruiter. If you can show how you can amplify the culture, it may be a deciding factor on why you end up receiving the job.

- **Technical Needs**: This is necessary. If you can't show that you have a solid understanding of the area where the team is looking for support, you will not get a job.

- **Industry**: If you are applying for a job in the financial industry, at a minimum, you should read up on the industry. In the best case, you already work in the industry.

- **Second-tier responsibilities**: These are the teams you will be interacting with that require additional knowledge outside of what is required to directly do your job.

In the next section, we will discuss the career-wrecking consequences poor communication with recruiters can have.

Avoid neglecting to respond to recruiters after you have applied for a position

Failure to respond.

While working in the oil and gas industry, I had an initial interview with a recruiter, after which I took off for a week of work-related travel, followed by a week-long vacation.

During my travel and vacation, I neglected to check my emails and missed two calls from the recruiter. When I finally came back to reality, I got back to the recruiter and learned I was no longer being considered for the role. Honestly, I had forgotten about it, but it did teach me a lesson I hope to pass on to you:

Always have a communication strategy when applying for jobs.

In my example, I could have set up an auto-reply that stated I would not be able to respond to emails until I returned from my travels. This way, the recruiter would have been more likely to grant me additional time to respond. Better, inform the recruiter of upcoming work and personal plans you have coming up in the short term so that any confusion can be avoided. After interviewing, you decide that you are no longer interested in a job, do the polite thing, and inform the recruiter you are no longer interested in preserving your relationship for opportunities that may arise down the road. When applying for jobs, it is important to follow the **LinkedIn, Email, and Phone (LEP)** protocol, as shown in the following diagram:

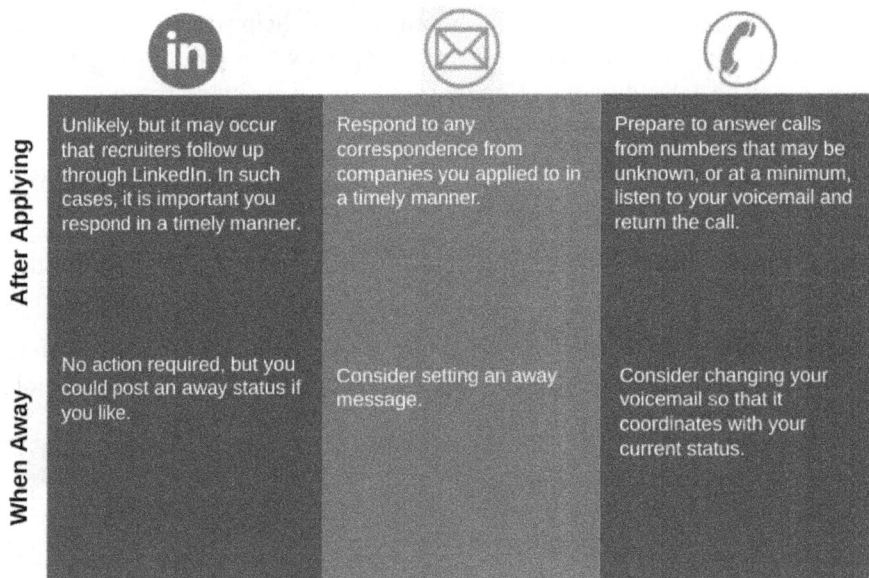

	in	✉	☎
After Applying	Unlikely, but it may occur that recruiters follow up through LinkedIn. In such cases, it is important you respond in a timely manner.	Respond to any correspondence from companies you applied to in a timely manner.	Prepare to answer calls from numbers that may be unknown, or at a minimum, listen to your voicemail and return the call.
When Away	No action required, but you could post an away status if you like.	Consider setting an away message.	Consider changing your voicemail so that it coordinates with your current status.

Figure 10.5 – LinkedIn, email, and phone follow-up protocol

The preceding diagram shows the practices that can increase the chances for success after initial contact with a recruiter. Next, we will cover the importance of consistent information across platforms.

Avoid inconsistent information across social profiles and your resume

When on LinkedIn, you will run across profiles of people who have highly exaggerated job titles, appear to be skilled at everything, and have rudimentary fluency in six languages:

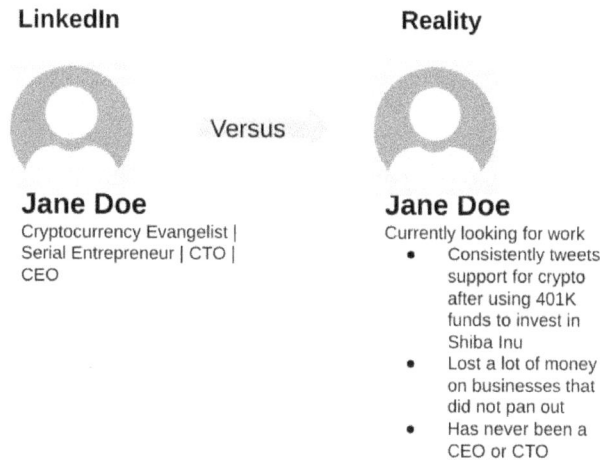

LinkedIn   **Reality**

Versus

Jane Doe   **Jane Doe**
Cryptocurrency Evangelist | Currently looking for work
Serial Entrepreneur | CTO | • Consistently tweets
CEO   support for crypto
                                              after using 401K
                                              funds to invest in
                                              Shiba Inu
                                          • Lost a lot of money
                                              on businesses that
                                              did not pan out
                                          • Has never been a
                                              CEO or CTO

Figure 10.6 – LinkedIn versus reality

The preceding example is meant to be an exaggerated version of what you will run across, but it also should cheer you on and tell you to keep your head up – most people are not as well off as their social profiles state. Exaggerating yourself on LinkedIn and other social profiles can adversely affect you, especially if you are currently applying for jobs.

Exaggerated information will get you noticed but once the full story is discovered, your reputation will take a hit.

So far, we have discussed things you should avoid doing to increase your chances of a positive outcome from an interview. Next, we will cover things you should do during the interview process to increase your chances of landing a job.

Things to do during the interview process

Throughout the many interviews I have been part of, both as the interviewer and the interviewee, I have come across several things that had a surprisingly positive effect on the process. In this section, I will cover several of these.

Discuss your side projects

Not all side projects hold weight when you are interviewing for a DevOps position; your rock collection holds little importance in an interview, but using a Raspberry Pi to create an AI that can distinguish between distinct types of rocks is cool and should be brought up. Use the discussion and your judgment to determine if your project holds significance.

You're interested in home automation? Me too!

I was in an interview for a lead engineering position and had made it through the initial interview with the hiring manager, as well as the technical interview. During a final follow-up interview, I was told it was down to me and one other individual. The conversation was more laid back than the previous discussions; the hiring manager told me he had been working on upgrading his cabin's internet connection so that it could handle more sophisticated home automation.

I took this opportunity to start discussing the home automation project I was working on. I do not believe this was the deciding factor that landed me the job; however, it did not hurt my chances either.

I was able to discuss topics such as wireless topologies and the C programming language because I mentioned my home automation side project, things that would have otherwise gone unnoticed as they were not part of my resume or things that I worked on in my professional work. It becomes even more important to discuss your side projects if you are looking for your first job in the DevOps field and have little real-world experience. Discussing your side projects will show the interviewer that you are excited about DevOps and excited about learning new things.

Come prepared, ready to discuss tool alternatives

DevOps tools come, DevOps tools go – that is the reality of it. If you have experience using GitHub and a company uses GitLab, do not worry about it! Be prepared to discuss similarities and differences between the two; show the interviewers you have done your homework and read up on the tool. The following table shows tools that can be used interchangeably during the interview process:

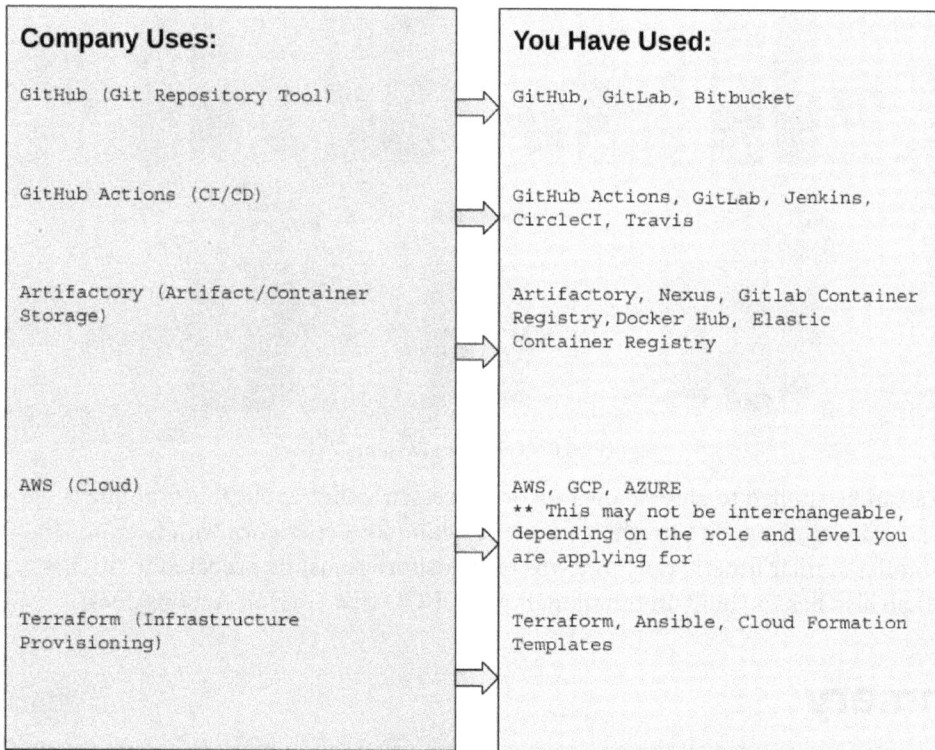

Company Uses:	You Have Used:
GitHub (Git Repository Tool)	GitHub, GitLab, Bitbucket
GitHub Actions (CI/CD)	GitHub Actions, GitLab, Jenkins, CircleCI, Travis
Artifactory (Artifact/Container Storage)	Artifactory, Nexus, Gitlab Container Registry, Docker Hub, Elastic Container Registry
AWS (Cloud)	AWS, GCP, AZURE ** This may not be interchangeable, depending on the role and level you are applying for
Terraform (Infrastructure Provisioning)	Terraform, Ansible, Cloud Formation Templates

Figure 10.7 – Tool alternatives

Sometimes, companies have hard tool requirements for candidates, but more times than not, if you have used a comparable tool, the knowledge will be seen as transferable. Cloud providers have a diverse collection of tools and require vast knowledge as well as certifications in some cases, which is why the skill sets may not be seen as transferable.

If the company is open to interviewing you for a role where you will be working with Azure and you have only worked with AWS, make sure that you come prepared to discuss how the tools are related between the two providers. A useful resource in an instance like this would be the following diagram, which was first discussed in *Chapter 3, Specialized Skills for Advanced DevOps Practitioners*:

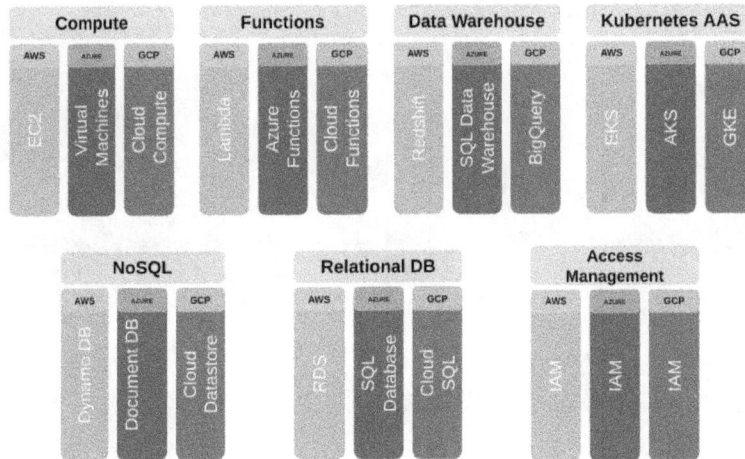

Figure 10.8 – Cloud equivalents

This idea can be applied to almost any tool or process; another applicable example is having experience using GitLab when the job is asking for experience with GitHub. The two both offer similar functionally in terms of repositories; just be prepared to discuss how GitLab also has the built-in functionality of CI/CD that GitHub Actions does.

Summary

In this chapter, we covered all the necessary changes you need to make to ensure you are presenting your best self to potential employers. We discussed the importance of having a complete and professionally written LinkedIn profile, as well as giving suggestions on how to improve the likelihood of being noticed by potential employers. Next, we covered how to update your resume in such a way that essential information is seen quickly both by automated systems and humans. Then, we covered the importance of having a personal web page, went through a tutorial on how to create a Hexo web page on GitLab, and sections that should be included in your arsenal web page. Finally, we covered other social sites that are not required but can increase the likelihood of you getting noticed by recruiters.

In the next chapter, we will discuss the importance of networking and how to do so on LinkedIn and at conferences.

11
Interviews with DevOps Practitioners

If you have followed along throughout this book, I am sorry. Honestly, I hope you have been able to gather a lot of valuable information from this book. This concluding chapter is based on conversations I have had with actual colleagues, as well as people I have interviewed.

In this chapter, we will cover the following topics:

- Interview with a senior DevOps manager
- Interview with a senior DevOps engineer
- Interview with DevOps architect consultant
- Interview with a tech executive passionate about neurodiversity and inclusion

Reading through the perspectives of individuals with varying degrees of experience will allow you to form a more holistic picture of what to expect as you look for your next DevOps position or try to land your first role as a DevOps engineer.

Interview with a senior DevOps manager

The first person who I interviewed also happens to be the co-author of this book, *John Knight*. John started with high hopes of becoming a game designer. 20 years later, he is a distinguished engineering leader:

John Knight
Director of Technology with a deep understanding of DevOps and Agile methodologies with a proven ability to lead diverse engineering teams.

Over 15 years of software development experience spanning 7 Fortune 500 companies and multiple consultancy firms.

Figure 11.1 – John Knight's bio

Reporter: Hi John, thanks for agreeing to meet with me this afternoon to discuss your career as a DevOps leader!

John Knight: No problem.

Reporter: To begin with, please tell us a little about yourself.

John Knight: I am John Knight, and I am an engineering manager with 16 years of experience in DevOps, 9 years of experience in the cloud, and 5 years of experience in team leadership. I have worked for and consulted with eight Fortune 500 companies and hold multiple certifications in the three big public cloud providers. In my free time, I collect master's degrees.

Reporter: Sounds like a successful career! Before moving on, can you touch on your statement about collecting master's degrees?

John Knight: I currently have three master's degrees and am working on a fourth. I complain that I do not have a lot of time, but I can never allow myself to not be in school; I am a knowledge nerd.

Reporter: Interesting; we will come back to that, but right now, can you tell us about how you ended up getting into DevOps?

John Knight: I was a game developer, or was trying to be. I worked on two independent MMOs before being recruited out of game development. Turns out the skills needed to deploy and patch games were also the same skills that are used to deploy and patch software. That is how I got into this field. The lead architect was too busy to do the deployments himself! 16 years later and I still thank him.

Reporter: 16 years, wow. What are some of the biggest changes in the field you have encountered and how have you ensured you stayed relevant?

John Knight: Technologies I used 10 years ago are mostly irrelevant now. You must keep up with the tech as it evolves and seek out knowledge from the experts, from the innovators, so that you exercise continuous improvement. Being a continual learner helps.

Reporter: You talked about being a collector of master's degrees, and how being a continual learner helps; I am starting to see a pattern. Can you expand on the importance of being a continual learner in the field of DevOps and how it has impacted your career?

John Knight: The technology I used 15 years ago, and some of what I used 10 years ago, is mostly obsolete. Someone coming in fresh would not have to survey the last 15 years of the field. The last 5 would be enough to be competent and even less in some areas. Being a continuous learner allows you, if nothing else, to be continuously upgrading your skills. It also allows you to be more creative in general, especially if you learn about areas outside your field of interest. That is the path that leads to true leaps – when you can tie formerly unrelated areas together. It is hard now because there is so much information that you cannot know or learn everything. So, being a regular learner allows you to learn in increments and makes it much easier to do. In terms of impact, I have a broader background than other similar candidates, so that gives me a competitive advantage if nothing else. Anything you can do to differentiate yourself by gaining more knowledge, credentials, and so on helps you in the long term. Knowledge compounds.

Reporter: Were there any times you encountered any setbacks or unplanned changes in your career?

John Knight: As far as setbacks or unplanned changes go, it always happens on a dynamic playing field. Once upon a time, I mostly worked in Windows systems and suddenly I moved to the Linux world. At some point, containers were introduced, as well as infrastructure as code. You must adapt or stagnate. Swim or sink.

Reporter: Wow – there is a lot of valuable information packed into that statement. Next, I would like to transition to how, when, and why you decided to transition into leadership versus continuing down the individual contributor path.

John Knight: It can be hard to transition into leadership because everyone wants experience, but there must be that first opportunity where you start with no experience, and it takes faith to give you that opportunity. In my case, I wanted to try my hand for several reasons. One is that I saw it as a logical career progression, to more opportunities in my career. Second, I was inspired by good and bad managers and wanted to employ what I learned and do things differently in areas where I thought I could do better.

The focus on learning, training, and improving your professional standing is also something that I encourage in all my team members, and you can see a vast improvement between the moment we started working together and a year later.

There are still many areas where I need to improve, but I feel that improvement in these areas benefits me in all areas of life and business, so it is a win-win.

In terms of how to make the transition, I started working in consulting in order to work with high-profile companies without needing to join their staff permanently. This gave me an outsider's perspective on their business and strengthened my people skills because they were the client and thus, I was now client-facing.

After doing that for a few years, I joined a small company and managed their DevOps team and initiatives. I found that the most important skill as a consultant after technical knowledge is to influence others. When you do not have authority, you must use diplomacy, charisma, and negotiation skills to get what you want. Learning this skill before you are a manager allows you to be a better leader.

Reporter: Influencing others – I see that as important in my role as a more senior individual contributor as well. What other skills are transferable, and what skills are completely different?

John Knight: The ability to mentor others and make people around you better. An experienced engineer is a catalyst for junior engineers and can help them get to where they are much faster. In terms of management, being able to prioritize work and analyze competing priorities is a key strength, but virtually anyone would benefit from strength in that area. Leadership is different from management. Leadership is owning things and being accountable. Even junior associates should strive to be leaders.

In management, you must negotiate a lot; you must exercise diplomacy to avoid scenarios where you win the battle but lose the war. There's a lot of managing people's expectations across multiple teams. As an individual contributor, you tend to focus on your immediate team.

Reporter: What is your advice to a junior engineer who is looking for a mentor?

John Knight: I think we should always try to model our behavior after the positive examples we see. That way, we can improve and be influenced by those around us that have something we do not or excel in an area. Having strong relationships with senior members of the team also opens potential mentorship opportunities.

You can also ask your immediate supervisor to play an active role in your career. Mentors can be found everywhere.

Reporter: A question many of us have for you revolves around the process of hiring; could you start by telling us about a few times where a candidate just did not deliver during the interview process?

John Knight: Usually, when I interview in screening, I look for experience, culture, and team fit, as well as softer people skills. I want people that work well with others. I have had people focus too much on compensation, overstate their skills, or show a lack of enthusiasm.

One time, I had two candidates and I sent both a curveball; a last-minute questionnaire for both to fill out. One returned it the next day, with a positive attitude. The other one was upset because this last hurdle was unexpected. The first person got the job.

I have been fortunate with my hires, and I am still updating my interview process.

Reporter: John, can you give us a deeper look into other technical evaluation exercises you have used in the past? Also, which ones did you feel gave the best insight into a candidate and what are you looking for as you review the technical challenges?

John Knight: So, there are many ways to evaluate technical expertise. Usually, I focus on experience. What problems did you solve with technology you claim you know? I usually assume that what they are telling me is accurate and then just look for experience. This allows me to evaluate multiple dimensions at the same time. Sometimes, you get character insights, such as if someone is great with collaboration versus a lone wolf, and so on.

Traditional metrics for technical acumen are useful but I usually leave that to other interviewers. Programming proficiency, systems experience, cloud expertise – these are all areas where it is easy to test general knowledge. I do not want to know if you can define something. You can look up every definition. I want to know what you can create by yourself, and what the limits of your knowledge and skills are. I specifically look for complements to my existing team, so having more than one skill set is always great.

Reporter: Thanks John, just a few more questions. Looking back at your career, are there any specific things you wish you would have done differently or avoided?

John Knight: I moved areas based on compensation needs and not based on what I wanted to do, or whether I was particularly interested in what the company was doing. I would advise against doing this, although I understand finances have a key role in every decision. Sometimes, I wish I had been brave enough to take a lower-paying job with a company or product I was passionate about. Regardless, my choices have resulted in a good career and outcomes, so I am very fortunate.

Other things I could have done differently include scanning for cultural fit more before taking the job. I have worked in a few places with hostile work environments and while no question or interview is full proof, as a candidate, you should also conduct your due diligence on the company and make sure their practices align with your beliefs.

Reporter: Final question, John – I have read some of your works on AI and future technologies; do you feel AI will affect the field of DevOps in the future? What should individuals do to prepare for the future of DevOps?

John Knight: Not really AI, but **AI-assisted automation** and **machine learning**. There are two intersections. One involves using DevOps to deploy AI or ML into production. This is known as operationalizing AI/ML or AI/MLOps.

The other involves using AI agents to respond to events automatically, or ML to make predictions on things to happen or to make recommendations. This field is ripe for innovation. I would think agents that automatically remediate services, such as a **Site Reliability Engineer** (**SRE**) agent, should be coming soon if they do not exist already somewhere; intelligent testing automation, intelligent resource allocation, AI/ML in threat prevention, detection, remediation, and so on.

To prepare, you must broaden your horizons and continuously learn. Cloud providers are great at teaching you how to use their services, but you must understand the underlying knowledge to maximize your utilization. Thankfully, we live in an age where learning resources are abundant, where we can take classes online at the best universities in the world, and where we can acquire every kind of information in the blink of an eye at the lowest possible costs. Truthfully, if you have access to the internet, there is nothing you cannot begin to learn. It is an amazing time to live in.

Reporter: It sure is, John. Thanks for taking the time to share your insights with me today and I look forward to doing it again.

Interview with a senior DevOps engineer

The second individual I interviewed is *Veeral Patel*. Veeral is an individual who I had the opportunity to hire into his first role as a DevOps engineer:

Veeral Patel

Currently working to connect top tech talent with their ideal job working with great teams and cutting edge technologies at Fannie Mae.

I understand top talent plays a crucial role in any business large and small.

Former policy nut and teacher who loves all things antique and chipped who writes and runs in her spare time.

Figure 11.2 – Veeral Patel's bio

Reporter: Good afternoon, Veeral, thanks for agreeing to sit down and discuss your journey to becoming a DevOps engineer.

Veeral Patel: Thanks Nate, no problem!

Reporter: Before we get started, can you share some background information about yourself?

Veeral Patel: Sure! So, I have an educational background in software engineering and data science. In the real world, I have worked in tech for the past 4 years. In terms of technical experience, I have done full stack development with React and C#, scripting with Python, and done DevOps automation work in Jenkins, GitLab, and Azure DevOps.

Reporter: When did you discover your passion for technology and what was the defining moment where you knew you wanted to write code?

Veeral Patel: I knew I wanted to write code when I took an introduction to a computer science course in college, and it was the only class I would attend regularly, even though it was an 8 A.M. class in my first year.

Reporter: Haha, so you knew it was your calling when it was worth putting effort into it; I love it! While in college, did you end up landing any internships? If so, what were they, and would you mind going into detail on what your responsibilities were and what you learned? Also, did your internship in any way influence the classes you took or subjects you studied post-internship?

Veeral Patel: I did – I interned at *Medicon Health*! When I was there, I worked on software testing, we were doing integration testing and I used this tool called **Applitools**. That is when I learned how to find elements from a web browser and write scripts for clicking on elements and doing data validation. The internship did not affect the classes I took since I took the required courses that are needed to complete my degree as quickly as possible.

Reporter: Do you feel your internship affected your ability to get a job out of college? Also, what was your first job post graduating college? Did you pursue a secondary degree directly after getting your undergraduate degree?

Veeral Patel: It was easy getting a job after college. I had several offers before graduation, so I went with the company offering a higher salary. I waited a year out of college to pursue a secondary degree, which allowed my employer to pay for a small portion of it.

Reporter: Sounds like a financially smart decision! What are three pieces of advice you would give to software engineering students who will be looking for a job soon?

Veeral Patel: First, I encourage students to continue to learn modern technologies – the world of tech changes so much that you need to take time out of your workday to spend time learning.

Second, if you would not pursue a job if it paid half the amount, you are not that passionate about it. It's hard to make it far in tech without passion; there is just so much competition and everyone is smart.

Finally, do not be afraid to quit your job if you are not happy – there are too many opportunities out there to be stuck.

Reporter: That is solid advice that everyone would benefit from if they were to follow. I would like to switch gears and discuss how you transitioned from software engineering into cloud DevOps engineering.

Veeral Patel: I transferred from software engineering because I liked to be able to own certain pipelines or projects or initiatives. As a software engineer, I was part of a team that maintained or created an application; it was a giant cog. In DevOps, I get to work on certain projects and own them. Also, I get to work on so much innovative technology. For example, in DevOps, I get to touch things such as AWS, Terraform, Ansible, and so on. In software engineering, I could be stuck working on Java for years.

Reporter: We should put that on a poster – *Get into DevOps so you are not stuck doing Java for years*. That sounds like the reason I got into DevOps – I wanted to work on cool technology. When you first interviewed for a DevOps role on my team, your desire to learn and soak up knowledge was the reason you beat out other candidates with similar skills; can you tell me other areas in your career where this trait has paid dividends?

Veeral Patel: This ability and desire to learn and soak up knowledge has paid off in school as well; I have always thought formal education is very important and I think the fact that I have a few degrees can be attributed to my desire to learn. I can kind of talk about all sorts of technologies because I have spent some time at least looking into them, so interviewing is a lot easier, and I am never worried about volunteering for a task I know little about.

Reporter: Veeral, will you describe the perfect member you would want to hire to join your team? What qualities should they have?

Veeral Patel: The perfect member is a member that loves sharing ideas and does not mind taking on new challenges. I always like to hear other perspectives, and that alone can get the ideas flowing. Being technically adept is good but having a willingness to learn is just as important.

Reporter: You also have an advanced degree in data science. How does this knowledge complement your traditional DevOps skills? Do you have specific examples?

Veeral Patel: Having a degree in data science opens ideas and thoughts about what is possible with pipelines and moving data. I always think to myself about how some data or how some pipeline would be valuable in analytics or what could be done with data that currently is not readily available. When I worked in healthcare, I always thought about how much untapped data there is that is just going to the database and being displayed back to a user interface without any real analysis being done on it.

Reporter: Thanks for your time, Veeral – I look forward to speaking with you again.

Interview with a DevOps architect consultant

Some people have an impact on you. *Chris Timberlake* is one of those people for me. From the first time I attended a meeting with him, I could tell he was going to be someone that I was going to be able to soak knowledge from. This proved true during the time our paths crossed while we worked on an implementation for GitLab together:

Chris Timberlake

I'm currently a DevOps Consultant, but I've held many titles and roles throughout my career. I have a wide-range of experience, from Infrastructure, to Embedded Software Development, to Video Game Design and Development.

I have shipped numerous applications that serve hundreds, if not thousands, of customers a second, I've also worked on products that have never seen the light of day. Resolving an exception, incident, or unique challenge may be anxiety inducing; for me it's just a Monday.

Figure 11.3 – Chris Timberlake's bio

Reporter: Thanks, Chris, for agreeing to discuss your DevOps journey with me. Before we get started, can you introduce yourself and the experiences that have led you to where you are now?

Chris Timberlake: Thanks for having me, Nate! I started working with computers at a young age – before I was 10 years old. I started out working on GeoCities websites and booting tools for Yahoo Messenger, and then moved into programming video games based on the Half-Life Source Engine. These events were my introduction to programming. I have been addicted ever since. I took a short break to pursue a career in law enforcement. Then, I came back to programming and found myself through a series of events working for Red Hat on massive digital transformations. Now, I work at GitLab as a professional services architect, and I help lead digital transformations.

Reporter: Wait – so you are a technologist and an officer of the law? Impressive! Do you consider yourself a DevOps engineer? Can you explain how you differentiate between a DevOps engineer and a software engineer or are they the same?

Chris Timberlake: I do consider myself a DevOps engineer, but I also consider myself a half-decent software engineer and SysAdmin. I move between the separate roles very fluently. Previously, in 2015, I would say that a DevOps engineer was just a software engineer with SysAdmins abilities. However, the DevOps field has grown substantially and evolved with the cloud market. Today, I consider a DevOps engineer its own role. Today, a DevOps engineer is an engineer who handles most of the piping for automation and delivery.

Reporter: Chris, you have seen many sides of engineering and DevOps. Can you explain the biggest difference between working in DevOps as a consultant versus directly for a company?

Chris Timberlake: The biggest takeaway is your role. As a consultant, you must have an entirely different mindset and you have different responsibilities. As a consultant, you are supposed to be a leader of the field, an educator, someone people look up to for answers and to get out of a hairy situation. You also have to coordinate travel, expenses, being on call, and working in different environments – sometimes without the tools necessary. As a consultant, any role you have is much more difficult. For example, as an employee, you do not have to worry so much about ensuring you are on a plane in Atlanta at 6 P.M. to meet a connection in Dallas at 10 P.M. so that you can be at a meeting in San Francisco at 8 A.M.

As an employee, you have far fewer concerns. I think of being a consultant for DevOps as being two distinct roles in one. On the one hand, you are an individual contributing to DevOps engineering. On the other hand, you are a traveling leader or teacher who educates others on the topic.

Reporter: Thanks, Chris. As a follow-up, question do you think a consulting role is suitable for someone looking to start their career in DevOps or is it a role for experts only?

Chris Timberlake: It is suitable for either scenario. It is less about knowledge and more about the determination and ability to remain calm under pressure. As I started my consulting career, I certainly was not an expert on all things. Even today, I am asked about things I am not an expert on. But when faced with uncertainty, I make sure I go out and find those answers for folks. I make sure I educate myself on the subject at hand while keeping a clear mind about the topics. One of the things being a consultant has taught me is that I do not understand a solution or topic unless I can reliably, accurately, and properly argue both for and against that item. Being a consultant can be a stressful job. It is certainly not for everyone. It is less about what you know and more about how you work, being able to handle the stress of the role, and even actually enjoying the role.

Reporter: Thanks, Chris, I feel our readers will find this particularly useful as they navigate their DevOps career. I would like to switch gears and talk about open source. At least two of the companies you have worked for are considered pioneers in open source software – GitLab and Red Hat. Many would say they are also dominant players in the field of DevOps. Can you share your thoughts on the relationship between open source and DevOps?

Chris Timberlake: I mean, it is no secret that the future of software is open source. Historically, many of the DevOps tools and patterns came from open source projects. People would self-organize and work on a project together, then bam – you have a new product that is used by many and even corporations. Some may feel that open source is beneficial because it allows folks to be involved in software who may not otherwise be.

The true power of open source is transparency, collaboration, and, most importantly, open conflict that happens. All these things are key to producing something great. Red Hat is not just a company built on open source; they bring those values into how they work and operate. They allow employees to be themselves and are transparent with those employees about things. Red Hat was able to become the biggest acquired tech company to date at 34 billion dollars. GitLab also takes these values to heart – not just with its software but in the way it works. This resulted in a massive **IPO (Initial Public Offering)** this year.

Those same values that open source advocates and encourages drove those two companies to their heights and I believe will cement DevOps as a requirement for all software and IT companies in the future. This is because DevOps also encourages and implements those same values.

Reporter: I could not agree more! Now, going on a tangent to what you said about DevOps values, what do you feel are the most in-demand and desired soft skills for DevOps professionals and those looking to break into the field?

Chris Timberlake: I think no matter the role you are in, you should aspire to be a leader. Now, that does not mean you have to be a manager or even lead anything. But having leadership-like attributes will help you in your career goals anywhere. Having effective communication skills is necessary. Clear communication that is informative and receptive cannot go overstated. Having confidence is another important attribute.

Even if you must fake it till you make it. I see many amazing engineers trip themselves up with imposter syndrome. We are all here trying to build something great, we will all run into issues, and we will all fail. Lastly, I think accepting failure, being willing to risk failure, and then overcoming failure is important. Every engineer with strong soft skills has a story of when they failed hard.

Me? I lost 4 million dollars in revenue in an afternoon for a company I was employed by. Being able to iterate and learn from those failures, prevent them from happening, and build something better after a failure is very important. That's because everyone in any career will fail; it is what you do after you fail that is important.

Reporter: I would be interested to hear more on the story of how you lost $4,000,000 if you can discuss that. What happened and what did you have to do to remediate the situation? Also, were there any repercussions from your employer when this happened, and what was their response?

Chris Timberlake: Unfortunately, I cannot. There are privacy concerns. What I will say is what happened; I oversaw the release of some software to a shopping cart. It caused the shopping cart to not entirely break but have a reduction in function. So, we stopped many customers from checking out, but not enough to trip automatic alerts and monitoring.

As for repercussions, absolutely. I had to write a detailed document explaining how everything transpired., digging deep using the 5 Whys method. Then, I had to write a statement to the shareholders of the company as to what happened and assure them it would not happen again. All I will say is that JavaScript is very cool until there is a small syntax error that causes big problems.

Reporter: I understand. I understand outside of work, your hobbies also include many software-related endeavors. Can you talk about these, and do your endeavors outside of work play into your success at work?

Chris Timberlake: Absolutely! I will not say my outside work successes have helped my career. But my failures have! I have done a significant amount of mobile, game, and desktop development, which is a fancy way of saying I have gone down the rabbit hole of finding issues with building toolchains and third-party libraries. It also means I have had experience with many technologies my normal job does not expose me to either.

A good example is when Apple flipped the switch on mobile apps and blocked any apps from being able to compile code on the fly. It broke numerous MongoDB libraries: being able to navigate and resolve this issue was a terrible endeavor that I learned a lot from. I use those lessons learned in my day-to-day work.

Then, there are the more fun rabbit hole projects, such as converting Quake 2 from C into C++ compilers on a weekend. This taught me the inner workings of some of the C and C++ compilers. Not all my projects involve software. I am also a huge car fanatic. Being able to diagnose and problem solve weird electrical issues with a car or engine helps me solve problems at work. These lessons are translated into software.

Reporter: I will not keep you too much longer, but I have a few more questions for you. What advice do you have to offer to individuals who want to get into the DevOps field?

Chris Timberlake: I would say that if you want to be in DevOps, you should spend a lot of time on side projects and build up a lot of knowledge. Your job will not expose you to everything you need to learn; you must be passionate about the field. Build a mobile app, make a video game, and then automate it.

Reporter: Last question – technology is changing so fast. Can you paint a mental picture of what the DevOps landscape will look like 20 years from now?

Chris Timberlake: In the future, I think we will see two things. First, we are going to see a consolidation of companies in DevOps as a fight for sales and market share increases. We have seen a tremendous consolidation over COVID.

With that, we are also going to see a lot of new companies start, where someone has a new process or idea, and they give it a shot and build a company from it. This is similar to how GitLab and many new tech IPOs started. Both of those are near term; 5-10 years out. I could not even guess 20 years out. Everything is moving so quickly that in 20 years we could have wearable VR and Star Trek holograms. Or, those things could be like flying cars, forever just an arm's length away.

No matter what, we are going to see increased transparency, collaboration, and conflict in software and DevOps. No matter what, we will be better off.

Reporter: Thanks for your time, Chris, I look forward to the next time we discuss all things DevOps.

Interview with a tech executive passionate about neurodiversity and inclusion

The next individual I interviewed was *Magnus Hedemark*:

Magnus Hedemark

I do a lot of side work in Neurodiversity Inclusion and any role I consider will have to have room for me to continue to lead in that space.

If you're looking to book time with me to explore a mentoring relationship, you can do so on my Calendly here: https://calendly.com/magnus919

Figure 11.4 – Magnus Hedemark's bio

Reporter: Hi Magnus, I am excited you agreed to let me interview you about DevOps. Would you mind telling us about yourself, and your career progression in IT and DevOps?

Magnus Hedemark: I took a very unconventional route into IT and that could be a very long story by itself. But I think throughout my journey, one thing that always served me well was to try to look ahead and see how I could be useful ahead of the known need. I like figuring out how things work, so rather than starting on the developer side, I came upon the Ops side. And I always had a knack for figuring out how to automate big, hairy, audacious things so that I could focus on smarter work. That was not common in my field in my early career, not like it is now. But there were a couple of big inflection points in the industry where the industry itself changed and my career changed:

- The industry started moving to Agile. I was on an Ops team at IBM where some great engineers from Brazil were huddling around their work every day, and sort of socially swarming around it in a way you do not tend to see in enterprise-driven Agile programs. This was fascinating to me and was a real reveal of what was missing in my career development.

- I had a CTO pull me aside and challenge me to consider a career in leadership. I was not sold on the idea, but he committed to spending time with me to understand what great looks like, and this was missing. I try to be that person now for others who I see with leadership potential. But I think when I was younger and earlier in my career, information was power, and job security meant not having successors. These days, information wants to be free, and great leaders are constantly grooming potential successors.

It turned out that I enjoyed leadership even more than engineering (though I am still constantly building things for fun). I went from just being a *brilliant jerk* who preferred to hack in isolation, to what I hope more folks would consider an empathetic and supporting leader who places a high value on collaboration. The progression was as much personal growth as professional.

Today, I lead a little over 300 people with a broad-ranging product portfolio in the TechOps space.

Reporter: Wow, thanks, Magnus – that was an amazing response! I guess I should not be surprised; I follow you quite closely on LinkedIn and am wondering about the following:

- What advice do you have for young IT professionals looking to advance their career in a DevOps direction?

- Second, what advice do you have for more seasoned IT professionals who are considering a career in management?

Magnus Hedemark: For folks just starting, I do not think there is a boot camp that adequately prepares you. Consistently, the absolute best people I have worked with in this field, the ones who earn well and are in demand, have a couple of defining traits:

- They are very curious people, always learning new skills for personal enrichment, and then applying those skills in their work. Even in a senior leadership role, I dedicate about 10% of my annual bonus every year to leveling up my skills. That might be paying for books or coaching. This year, I am using it to underwrite some HomeLab upgrades for learning Kubernetes much more deeply. If you are just getting into DevOps, some foundational texts are more of the beginning of the journey... if you have not read these things yet, you are missing what a lot of people around you are understanding:

  - **The Phoenix Project**

  - **Continuous Delivery**

  - **The Agile Manifesto** (it will take you less time to read than this answer) but even more importantly than that, the *12 principles* behind it

- They are very empathetic people. I mentioned the *brilliant jerk* paradigm before and there is not much room for it now. Learning to be a good listener, to be curious instead of furious, goes a long way. Some people are gifted in this area. Others, like me, have to work harder at it. But DevOps is as much about the cultural impact as it is the cool technical skills. I think learning about the **Westrum model** is one of the first things I would call out to make sure you are cultivating a high-collaboration engineering culture. I also think it's important to stay aware of the cultural sensitivities of underrepresented communities in tech. *Guide to Allyship* is a good place to start. Just being aware and mindful – those strengths are introduced when we make sure to include people with different backgrounds and different ways of thinking. But as a matter of respect, too, we need to invest in understanding how to include them.

For people in an individual contributor role thinking about moving into management... where do I begin? I think if you are not doing well at the curious and empathetic points I made in the previous question, I would advise that we have already got too many in this field who are missing these key traits and it may not be the right path for you. It is important to understand that going from individual contributor to manager is not a *promotion*; it is more of a lateral move to a completely different career track. For example, just to level set, where I work now, the highest-ranking individual contributor that can report to a senior manager is a senior staff software engineer. Even though there is a supervisory relationship between the two, they are at the same pay grade. When someone moves from senior staff software engineer to senior manager (a not-uncommon move), there is initially some disappointment that they are not getting paid more. Even as you move up in management, you may have highly skilled individual contributors earning more money than you. I want to tell you now and keep it filed away for when this inevitably happens – *get over it*. The jump in compensation does not tend to happen while going from individual contributor to manager (within the same organization), but when you jump from manager to director (manager of managers of individual contributors). But here is the thing – you could stay in an individual contributor role and still get that sweet pay bump by getting a promotion to principal software engineer. And your reporting relationship will change, because now, you outrank your senior manager, and you will instead be reporting directly to their director.

So, it all comes back to this, I think.

If you are thinking about making the move, think long and hard about what your motivations are. And think about your level of commitment to this path. Do not try to dither with *I'm going to still write code half the time*. Reality check: you are not going to write much code anymore. And if you do, you are now putting other engineers into a very uncomfortable position where they must tell the boss their code sucks. Just get over it, get over yourself, and commit to being a great leader. If you just want to be *an engineer with a badge*, this is probably not the right career move for you.

For me, my motivation is to liberate latent potential (in myself, in others, in teams of people, in lines of business, their products, their customers, and the communities that they work in).

Reporter: You have over 300 people who report under your leadership, meaning you have seen your fair share of interviews, both good and bad. If you do not mind, I am sure the readers would be interested in what made the good and bad candidates stand out.

Magnus Hedemark: I think a lot of other seasoned interviewers will nod knowingly when I say that most of the time, you know in the first 5-10 minutes if that person is a slam-dunk or a hard-no. But I still like to take more time anyway, because sometimes, a *slam-dunk* can turn into a hard no when you get to know them more (but I do not think I have ever seen an interview go in the other direction). Again, I think a lot of empathy and understanding must be shown regarding there being many ways to be *normal* in this world versus just those who we identify with (affinity bias). So, the following are some of the key things I think about when I am talking to someone:

- Do they objectively have hard skills that will raise the bar for this team?

- Might they be a *culture add* to the team? (I do not interview for culture fit; that is how you get a monoculture.)

- I ask some questions to help a candidate expose if they are a high risk for breaking the *No Jerks Rule*, which I take very seriously as responsibility for cultivating great team cultures. You can't let even one toxic person into your team, and if you find them, you cannot let them remain.

Reporter: Thanks, Magnus, for your time. I enjoyed our conversation, and I am sure the readers will as well.

Summary

This chapter was a culmination of interviews with four DevOps professionals with varying levels of experience. We had the opportunity to speak with John Knight, who gave an insight into hiring practices, as well as what he looks for in terms of skills when hiring DevOps engineers. The DevOps leader also gave insight into how someone may transition from an individual contributor role to a manager. Lastly, John gave us his prediction for the future of DevOps.

Next, we had the opportunity of speaking with senior DevOps engineer Veeral Patel, who gave insight into how he landed a job in DevOps and gave helpful tips for students looking for a job.

Chris gave insight into the relationship between open source and DevOps. The major takeaway from Chris's interview is that DevOps is truly a mindset, and if you want to succeed, you cannot be afraid to fail.

Finally, we had the pleasure of speaking with Magnus, a technology executive who has a storied career in DevOps and was full of insight and knowledge for individuals looking to either get into the field of DevOps or transition into leadership.

Index

Packt>

Packt.com

Subscribe to our online digital library for full access to over 7,000 books and videos, as well as industry leading tools to help you plan your personal development and advance your career. For more information, please visit our website.

Why subscribe?

- Spend less time learning and more time coding with practical eBooks and Videos from over 4,000 industry professionals

- Improve your learning with Skill Plans built especially for you

- Get a free eBook or video every month

- Fully searchable for easy access to vital information

- Copy and paste, print, and bookmark content

Did you know that Packt offers eBook versions of every book published, with PDF and ePub files available? You can upgrade to the eBook version at packt.com and as a print book customer, you are entitled to a discount on the eBook copy. Get in touch with us at customercare@packtpub.com for more details.

At www.packt.com, you can also read a collection of free technical articles, sign up for a range of free newsletters, and receive exclusive discounts and offers on Packt books and eBooks.

Other Books You May Enjoy

If you enjoyed this book, you may be interested in these other books by Packt:

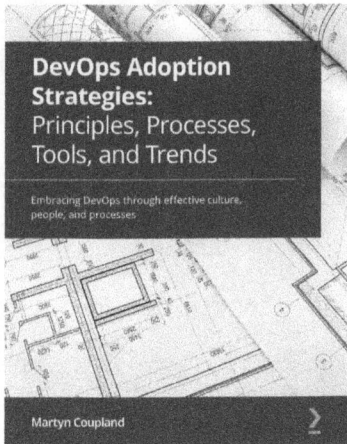

DevOps Adoption Strategies: Principles, Processes, Tools, and Trends

Martyn Coupland

ISBN: 9781801076326

- Understand the importance of culture in DevOps
- Build, foster, and develop a successful DevOps culture
- Discover how to implement a successful DevOps framework
- Measure and define the success of DevOps transformation
- Get to grips with techniques for continuous feedback and iterate process changes
- Discover the tooling used in different stages of the DevOps life cycle

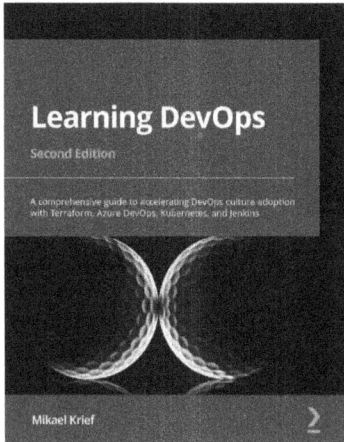

Learning DevOps - Second Edition

Mikael Krief

ISBN: 9781801818964

- Understand the basics of infrastructure as code patterns and practices

- Get an overview of Git command and Git flow

- Install and write Packer, Terraform, and Ansible code for provisioning and configuring cloud infrastructure based on Azure examples

- Use Vagrant to create a local development environment

- Containerize applications with Docker and Kubernetes

- Apply DevSecOps for testing compliance and securing DevOps infrastructure

- Build DevOps CI/CD pipelines with Jenkins, Azure Pipelines, and GitLab CI

- Explore blue-green deployment and DevOps practices for open sources projects

Packt is searching for authors like you

If you're interested in becoming an author for Packt, please visit `authors.packtpub.com` and apply today. We have worked with thousands of developers and tech professionals, just like you, to help them share their insight with the global tech community. You can make a general application, apply for a specific hot topic that we are recruiting an author for, or submit your own idea.

Share Your Thoughts

Now you've finished *The DevOps Career Handbook*, we'd love to hear your thoughts! Scan the QR code below to go straight to the Amazon review page for this book and share your feedback or leave a review on the site that you purchased it from.

`https://packt.link/r/1-803-23094-0`

Your review is important to us and the tech community and will help us make sure we're delivering excellent quality content.

www.ingramcontent.com/pod-product-compliance
Lightning Source LLC
Chambersburg PA
CBHW080532220326
41599CB00032B/6280